Environmental Pollution and the Media

This book offers a theoretically informed empirical investigation of national media reporting and political discourse on environmental issues in Australia, China and Japan. It illuminates the risks, harms and responsibilities associated with climate change through an analysis of pollution, adopting an interdisciplinary approach drawing on both the social sciences and humanities. A particular strength of the work is the detailed analysis of the data using a range of both quantitative and qualitative techniques, enabling the authors to reveal in rich and compelling detail the complex relationship between risk and responsibility in the climate change discourse.

The case studies of Australia, China and Japan are set in the current literature as well as in the historical context of climate change in these three countries. The analysis of the media discourse on the Great Barrier Reef in Australia demonstrates how the mining of coal for overseas markets has led to devastating harm to the life of the reef. A critical discussion of the Chinese documentary, *Under the Dome*, shows how this medium has played a crucial role in building awareness of the harm from atmospheric pollution among the citizens, shaping attitudes and promoting action. The first case study of Japan elucidates how cross-border atmospheric pollution from China forges a chain of responsibility for responding to climate change, running from the state to society. The other case study of Japan demonstrates how 'smart cities' have emerged as a way to mitigate the risks and harms of climate change. The Conclusion draws together the similarities and differences in how climate change is addressed in the three countries.

In all, *Environmental Pollution and the Media: Political Discourses of Risk and Responsibility in Australia, China and Japan* uncovers the dynamics of the triadic relationship among risk, harm and climate change in Australia, China and Japan. By so doing, the book makes an original and timely contribution to understanding comparative media, discourse and political debates on climate change.

Glenn D. Hook is Toshiba International Foundation Anniversary Research Professor in the School of East Asian Studies at the University of Sheffield, UK. He has published widely in Japanese as well as in English on Japanese and East Asian politics and international relations, especially in relation to security, risk and governance.

Libby Lester is Professor of Journalism, Media and Communications at the University of Tasmania, Australia. Her research on public debate over land use and environmental risks is widely published internationally.

Meng Ji is Associate Professor of Chinese Studies at the University of Sydney, Australia. She specialises in advanced digital humanities research methodologies, especially the design and the computational analysis and quantitative modelling of multilingual corpora ranging from classical literature and historical linguistics to global environmental media, cross-national health risk assessment and health references and guidelines for non-communicable diseases.

Kingsley Edney is Lecturer in Politics and International Relations of China in the School of Politics and International Studies at the University of Leeds, UK. He is the author of *The Globalization of Chinese Propaganda: International Power and Domestic Political Cohesion* (Palgrave Macmillan, 2014).

Chris G. Pope is a PhD student of the School of East Asian Studies, the University of Sheffield, UK, and part of the White Rose East Asian Centre. His thesis, which employs a mixed-methods approach to examine the speeches of Japanese Prime Minister Shinzō Abe, is funded by the Economics and Social Research Council.

Luli van der Does-Ishikawa is a research team member of the European Research Council project 'War Crimes and Empire' in the Faculty of Asian and Middle Eastern Studies, University of Cambridge, UK. She specialises in the interdisciplinary quantitative and qualitative study of sociopolitical discourses with focus on the transient loci of risk and responsibility in multistakeholder media discourse.

Routledge Studies in Environmental Communication and Media

A full list of titles in this series is available from www.routledge.com/sustainability/series/RSECM

Environmental Communication and Travel Journalism
Consumerism, Conflict and Concern
Lynette McGaurr

Environmental Ethics and Film
Pat Brereton

Environmental Crises in Central Asia
From Steppes to Seas, from Deserts to Glaciers
Edited by Eric Freedman and Mark Neuzil

Environmental Advertising in China and the USA
The Desire to Go Green
Xinghua Li

Public Perception of Climate Change
Policy and Communication
Bjoern Hagen

Environmental Communication and Community
Constructive and Destructive Dynamics of Social Transformation
Tarla Rai Peterson, Hanna Ljunggren Bergeå, Andrea M. Feldpausch-Parker and Kaisa Raitio

Environmental Communication Pedagogy and Practice
Edited by Tema Milstein, Mairi Pileggi and Eric Morgan

Environmental Pollution and the Media
Political Discourses of Risk and Responsibility in Australia, China and Japan
Glenn D. Hook, Libby Lester, Meng Ji, Kingsley Edney and Chris G. Pope, with contributions from Luli van der Does-Ishikawa

Environmental Pollution and the Media

Political Discourses of Risk
and Responsibility in Australia,
China and Japan

**Glenn D. Hook, Libby Lester,
Meng Ji, Kingsley Edney and
Chris G. Pope**

*with contributions from
Luli van der Does-Ishikawa*

Routledge
Taylor & Francis Group

LONDON AND NEW YORK

First published 2017
by Routledge

2 Park Square, Milton Park, Abingdon, Oxfordshire OX14 4RN
52 Vanderbilt Avenue, New York, NY 10017

Routledge is an imprint of the Taylor & Francis Group, an informa business

First issued in paperback 2019

British Library Cataloguing-in-Publication Data
A catalogue record for this book is available from the British Library

Library of Congress Cataloging-in-Publication Data
A catalog record for this book has been requested

ISBN: 978-0-415-71031-2 (hbk)
ISBN: 978-0-367-35054-3 (pbk)

Typeset in Galliard
by Apex CoVantage, LLC

Contents

List of figures   viii

List of tables   ix

Preface   x

Acknowledgements   xi

1   Introduction   1

2   Risk, responsibility and pollution in Australia, China and Japan: a regional overview   17

3   Transnational discourses of risk and responsibility from Australia: coal, pollution and the Great Barrier Reef   45

4   Responsibility and the risks of air pollution in China: media activism and the case of *Under the Dome*   74

5   Mediating risk communication and the shifting locus of responsibility: Japanese adaptation policy in response to cross-border atmospheric pollution   98
LULI VAN DER DOES-ISHIKAWA AND GLENN D. HOOK

6   Visions of a super smart society: risk management and responsibility as adaptation   146

7   Conclusion   188

Index   195

Figures

1.1	Key procedures of the empirical data analysis	10
1.2	Contrastive MSI on Type 1 ERMI – framing environmental health risk in the media	11
3.1	Structure of the Australian news data set used (in running tokens)	62
3.2	Great Barrier Reef reporting in Australia from 2011 to 2015	63
5.1	Number of publications on PM2.5, risk and responsibility	111
5.2	Correspondence analysis of the frequently occurring words over the nine periods	116
5.3	Co-occurrence of frequently occurring words in the risk and responsibility data – all nine periods	117
5.4	Annexation plot	121
5.5	Distinct word groups observed in multidimensional scaling (MDS)	122
5.6	Heatmap with the occurrence rates of the members of concept groups	124
5.7	Co-occurrence network of coding rules and the nine periods	126
6.1	Number of debates containing the term 'smart' from 1951 to 2015	155
6.2	Application of 'smart' during individual sessions in the Diet	156
6.3	Most frequently used 'smart' phrases in the Diet debates (2004–2015)	157

Tables

2.1   Japanese polls: 'What should be done about the future
       of nuclear power?'   35
3.1   Sample semantic analysis categories   61
5.1   Homogeneity of the text by types of news providers   107
5.2   Distribution of texts and their length over the nine periods   112
5.3   Coding rules and codes for risk and responsibility versus
       Action, Actor and Sources   123
5.4   English translation of coded words in alphabetical order   134
6.1   'Smart' categories compiled from National Diet Proceedings   171

Preface

This book is an experiment in translation. First coming together in 2014, our small team of authors reflected the diversity of many contemporary research efforts taking place around the world. We worked across three continents and represented six nationalities – Japanese, Chinese, British, Dutch, Australian and New Zealander – and had diverse disciplinary backgrounds, including political science, area studies, linguistics, translation studies and media studies. Our common language was English, but it was not everyone's native language. We drew on an array of methods and theories to study our range of interests – methods that had overlapping features but had variously evolved and been named and renamed to disguise common histories – and interests were siloed from each other. We published in different journals and participated in different conferences.

Could we find a common disciplinary language? Or, if not that, at least a shared problem which we could each express in our own terms and yet all understand? Our attempt to do so is reported in the following chapters. The responsibility for each chapter was: Chapter 1, Libby Lester and Meng Ji; Chapter 2, Glenn D. Hook, Libby Lester and Kingsley Edney; Chapter 3, Libby Lester and Meng Ji; Chapter 4, Kingsley Edney; Chapter 5, Luli van der Does-Ishikawa and Glenn D. Hook; Chapter 6, Chris G. Pope; and Chapter 7, Glenn D. Hook.

Acknowledgements

Our work was only possible with the support of the Toshiba International Foundation, which provided funding to bring together team members led by Glenn Hook for two workshops in 2015–2016. A part of chapter 3 is derived from an article published in *Environmental Communication* (2016, 6:791–802), available online: www.tandfonline.com/doi/pdf/10.1080/17524032.2015.112 7849. These were hosted by the International House of Japan, Tokyo, where we discussed the project and draft versions of our chapters. We would also like to thank the staff members at International House, whose impeccable service made our stay so memorable. The Toshiba International Foundation and the Australian Research Council (ARC) through its Discovery scheme provided support for a final symposium at the University of Tasmania.

Libby Lester would like to thank the ARC for contributing research assistance support to the project (DP150103454, 'Transnational Environmental Campaigns in the Australia-Asia Region'). Meng Ji has received support from the Worldwide Universities Network Research Development Fund ('Climate Change in the Media'), the Faculty Research Support from the University of Sydney and the ARC (DP150102405, 'A Cross-National and Cross-Cultural Study of Global Translation Industry'). Research and Innovation Services at the University of Leeds provided Kingsley Edney with funding for the early stages of the project. Chris Pope thanks Makoto Shirai and Mariko Kuwayama from the Toshiba International Foundation; Tomohiro Hirayama, Tadayuki Nakamura and colleagues from the Toshiba Group Initiative for the Environment for their presentations and discussions with the team; Jaegul Choo for his presentation regarding a new software in the field of text mining analysis; and Professor Etsushi Tanifuji of Waseda University for his support alongside that of his research students Aya Kudo, Yu Haichun, Rob Fahey, and Kentaro Nagai. Our final thanks to Janine Mikosza for her work on turning our drafts into a manuscript.

1   Introduction

This book represents research from a range of disciplines from the social sciences and humanities, six nationalities, divergent theoretical and methodological approaches and numerous spoken languages, although we use English here as our vehicle for communication. It is an exercise in scholarship under globalisation or, more precisely, an experiment in translation. As Silvio Waisbord has written of globalised research efforts in the field of communication:

> The question is whether academic interest in promoting dialogue across intellectual heteroglossia is as strong as the desire to live comfortably within homophonic academic tribes. If the way the field historically developed foreshadows the future, then globalization will continue to facilitate conversations among specialized academic cultures around the world, but it might not prompt wide enthusiasm in transcending difference. In a field brimming with academic diversity cultures, difference may be tolerated, yet toleration might not imply actual engagement with difference and the politics of translation.
>
> (2016: 881)

Here, Waisbord is not focused purely on linguistic translation – on 'semantics and the biases of language' (although this is always a factor) – but on the 'translatability of differences across intellectual traditions and the institutional logics of academe' (2016: 872). Such a focus, he suggests, reveals why the growth in cross-national comparative studies has not been accompanied by a similar trend across theoretical or methodological approaches. 'Translation', he writes, 'demands commonalities as well as a willingness to overcome differences' (2016: 879). He continues:

> Bridge crossing might be desirable, but it will not gain much traction as long as scholars are, at best, mildly interested in finding intellectual kinship across difference. Without kinship, there is no translation – no search for commonness among difference. Finding and nurturing intellectual kinship requires openness to others. A dialogue between self and stranger, translation requires mutual curiosity and the welcoming of differences . . . It

demands that participants are willing to engage with others and be open to mutual understanding. Interscholarly interest, cultural open-mindedness, and receptiveness to difference are basic conditions for translation across academic cultures.

(2016: 880)

Our group worked to these conditions. Teasing out scholarly definitions and meanings that are presumed knowledge within disciplines, and stripping down terms, methods and different approaches, we sought to find common ground: a shared problem and a translation. We had come together to examine the media and political communication of the risks posed by pollution and climate change. Whether we studied Japanese government policy, Chinese political activism or Australian media discourses, each of us had been confronted by these risks. But our commonalities went deeper. We were all interested in how the nexus of climate change, media and policy has become increasingly amorphous, often disconnected from its causes and impacts, as well as the responsibility to respond to them, at the governmental and societal level. Given that this is a crucial point for the formation and enactment of public opinion and decision-making, it is worthy of serious investigation. It is a problem that well illustrates why we need to cross bridges.

Numerous studies have now charted a rise and fall in interest and action on environmental risks, including climate change, with media acknowledged as playing a central role in how publics and policymakers understand and respond to these risks. Research has drawn attention to the ways in which the professional practices of journalists and the logics of news organisations inadvertently gave voice to powerful interests (Hutchins and Lester 2015) and scepticism and denial, specifically when 'balancing' news coverage on climate change (Boykoff and Boykoff 2007; Boykoff 2013) or negotiating scientific ideas of 'uncertainty' (Allan 2002; Pollack 2005). Media carriage or containment of spectacular and symbolic images and messages to create awareness of risks and suffering is now better understood (Boykoff and Goodman 2009; Lester and Cottle 2009; Doyle 2011; Hansen 2011), as is the existence, deployment and impacts of strategic political communications and framing within environmental media discourses (Carvalho 2007; Hulme 2009; Nisbett 2009; Anderson 2013). The relationship between public opinion on climate change, mediated awareness and political affiliations has also been explored, including in terms of trust (Leiserowitz *et al.* 2013), patriotism (Tranter and Lester 2015) and fear (O'Neill and Nicholson-Cole 2009).

International comparative work has provided insight into the dynamics of and differences within environmental coverage at a national level (Painter 2013; Schmidt *et al.* 2013; Schmidt and Schafer 2015), although this work still needs to confront the significant challenges posed by: (a) various media systems (Hallin and Mancini 2004); (b) dominant US and European paradigms, foci and/or personnel within the study of media and communications and mediated environmental conflict; and (c) looking behind texts to identify different cultural, professional and political practices that impact on media content (Lester 2015). Within the Asia-Pacific, particularly the Association of Southeast Asian Nations (ASEAN) and Pacific Island nations, the relationship between media, politics and the environment is particularly fraught, given the disproportionate impacts of a changing climate on poor and developing regions. Yet, here the relationship remains understudied when compared to the amount of research emerging from Europe and North America.

The transnational flow of political information about environmental risk is a current frontier for researchers. Studies of globalisation and cosmopolitanism suggest the existence of publics wanting to intervene in distant conflicts to alleviate distant problems and associated suffering (see, e.g., Beck 2006, 2009; Fraser 2007) and that media can provide a forum for exchange of opinions over political decision-making, legitimising environmental governance (Nanz and Steffek 2004). However, comparative and in-country empirical research in relation to environmental risks and the media can provide only limited insight into how these publics are formed and act. According to Hutchins and Lester:

> A level of depth and scope in empirical evidence is needed that is commensurate with the reality of a 'global, media age' (Cottle 2013), and it is this challenge that promises to advance the theorization and understanding of mediatized environmental conflict at this moment in world history. The conduct and potential implications of environmental conflict occurring across borders and between nations and regions has been well recognized. But extensive and detailed evidence of precisely how governments, industry, activists, and the news media respond to environmental issues that are transnationally manifest is lacking. In Australia, for example, the impacts of the mining and forestry industries are evident through trade and diplomatic relations with countries located throughout the Asia-Pacific region, including Malaysia, Indonesia, South Korea, China, India, Timor-Leste, and Papua New Guinea. These varied relations and their manifold environmental consequences require that media, political, and information flows in and between these nations and across the region are identified and evaluated. To study conflict in this way and at this scale is no small task, encompassing intricate networks of environmental concern, strategic webs of media and political influence, public policy debates, and bi- and multi-lateral trade negotiations and deals. Nonetheless, it is imperative that this research challenge is met, as this is the arena in which global environmental futures are set to be determined.
>
> (2015: 353)

In attempting to meet this challenge, our team has focused on analysing media discourses of pollution created by fossil fuels: a visible, breathable manifestation of climate change. In particular, we focus on pollution produced by burning coal. This 'pollution' is dug from the ground, transported in massive ships through dredged sea channels out into oceans, to distant locations, where it is burned for electricity and steel production. The burning of coal generates gases that change the composition of the atmosphere, affecting our climate and weather. It also blackens buildings and enters lungs. Even a gentle westerly wind will carry this pollution across national borders, forcing residents in cities separated by seas to don face masks or to stay indoors. International banks and trading companies, with headquarters and offices across the countries where this pollution is carried by winds and complex supply chains, finance its production. Individuals support it when they turn on an electric light, and condemn it when they check particle concentrations on their smartphones.

How, we ask, is responsibility for this pollution that carries transnationally and is created transnationally attributed within media discourses? The literature on responsibility attribution within media crosses a range of broad subjects, including disasters (Pantti *et al.* 2012; Ewart and McLean 2015), health (Holton *et al.* 2012; Zhang *et al.* 2015), sports doping (Starke and Flemming 2015) and poverty (Kim *et al.* 2010; Yua *et al.* 2015), among others. Those that focus on environmental concerns (see, e.g., Olausson 2009, 2011; Liang *et al.* 2014; Lester 2016) tend to have a foundation premise that these framing and attribution processes produce identifiable debates and outcomes in the social world. They consider 'responsibility' both in terms of attribution of blame and assigning expectations of response and action (Funtowicz and Strand 2011), and seek to identify the intersecting sectors, issues, processes and actors that actively construct the framing of responsibility and attribution of blame in media reporting.

Responsibility attribution, they suggest, is constructed through dynamic interactions between and across the following areas:

1   *Institutional decision-making and actions*: governments, politics, natural disaster prevention and planning, international forums and negotiations (e.g. climate change conferences);
2   *Symbolic power and discourse*: media representations, activist communication strategies;
3   *Scientific research*: research activities, findings, announcements, and the responses to these (e.g. the Intergovernmental Panel on Climate Change [IPCC]);
4   *Non-human actors*: the atmosphere, marine environments, wilderness, animal species;
5   *Industry activity*: resource allocation, production sites, labour relations, disaster management and clean-up, corporate communication and public relations (PR), political lobbying (e.g. agriculture and mining);
6   *Economics and the market*: corporate strategy, social licensing, trade negotiations and agreements; and,
7   *Social actors*: individualisation, health outcomes, representatives, citizen bodies, consumption decisions, audiences.

This book thus represents the collaborative output of experts from a range of disciplines, whose aim is to provide new insight into media roles in the communication of environmental harms and politics. We do this by focusing on discourses of risk and responsibility about atmospheric pollution within the Asia-Pacific region, concentrating on Australia, China and Japan. The next section summarises, analyses and evaluates the methodological challenges and innovations made in this book, illuminating how our joint efforts have attempted to contribute to developing new interdisciplinary research paradigms appropriate to the study of cross-national environmental discourse.

Methodological innovation

One of the key methodological innovations of this book is the use of automatic or semisupervised data coding and processing techniques in the study of media about the environment. Data mining methods provide an effective tool to uncover important patterns and trends in the media reporting of environmental issues. These data mining techniques based on detailed linguistic analysis have proven significantly more productive and versatile than traditional frequency-based analyses. Noteworthy features of the methodologies and analytical techniques developed and tested in this book include:

- The development of new analytical frameworks and the verification of theoretical frameworks adapted from cognate research fields such as international politics and national security;
- The testing of novel analytical concepts in environmental discourse, including new models for multisectoral interaction and partnership building on managing environmental health risks and growing environmental innovation businesses.

The following section illustrates the book's methodological advances by summarising and highlighting the salient features of the research methodologies and approaches developed in the chapters to follow. Of particular note is how this joint work has applied techniques from corpus linguistics and empirical language and cultural studies to enhance transdisciplinary collaboration to tackle research issues of contemporary social significance. Meanwhile, our combined methodologies have provided an opportunity to assess the significance of the research findings obtained in our country-focused case studies. The case studies are set within the general context of environmental risks and responsibilities, initiatives of environmental social innovation formulated and led by different social and political sectors and actors and the emerging multisectoral interaction and partnership building on environmental risk management and innovation.

These reflect some emerging and rapidly growing areas in the multidisciplinary and transnational study of environment and development. The wide methodological scope and approaches developed in this book are a timely contribution for research students and academics intending to develop new interdisciplinary approaches in environmental studies. As noted above, the case studies presented in this book on Australia, China and Japan have attempted to break down the methodological barriers between various disciplines from the social sciences and the humanities such as international politics and security, media and journalism, area studies (e.g. Japanese Studies and Chinese Studies), applied language studies, computational linguistics and textual statistics.

As part of its aims, our book uses linguistically orientated approaches to the processing and analysis of environmental media texts developed collaboratively by the authors. The latest research from applied linguistics underscores some key aspects and stages of the empirical cases presented in this volume. These include the collection of environmental media materials and sources, the design and development of data coding schemes to facilitate the automatic corpus annotation and information retrieval process and the automatic extraction and modelling of textual patterns in the data sets collected. All these represent the latest research developments in corpus linguistics and its ongoing outreach into the social sciences. This fits in with the general and increasingly salient research trend of large data–informed disciplinary transformation across a number of research areas in the natural sciences, social sciences and humanities, a momentum built up by the availability of ever larger amounts of information and data on the Internet. The interdisciplinary transformation witnessed in a number of research fields requires the development – normally through collaborative effort among experts from different fields – of research approaches and methodologies able to effectively process and extract significant and meaningful patterns from the large data or corpora constructed.

Overall, we have explored and illustrated a number of innovative approaches to the effective processing and analysis of large media data sets which may be broadly divided into two approaches to environmental media analysis: corpus data based and corpus data driven. Important methodological differences underlie these two approaches, which has stirred much debate in the corpus linguistics communities in the last few years (Hunston and Francis 2000; Laviosa 2002; Biber 2012). The corpus data–based approach has proven most efficacious when existing theoretical hypotheses or frameworks inform the analysis of particular data sets. The aim of using corpus data within this corpus data–based paradigm is therefore to verify the suitability and applicability of existing theoretical instruments in analysing and explaining the patterns extracted from purposely built databases.

Another set of corpus methodologies explored in this book is the corpus data–driven approach. This particular approach or methodological orientation is distinctly more exploratory and experimental in comparison to the corpus data–based approach described above. As its definition suggests, the development of analytical frameworks within this research paradigm is driven and backed by the discovery, modelling and retrieval of meaningful patterns in relevant data sets. The corpus data–driven approach has proven most efficacious when the aim of the research is to reveal novel patterns in the data sets or when there is a lack of relevant theoretical frameworks able to effectively support or guide the data analysis. We explore the feasibility, productivity and significance of this corpus data–driven methodology in order to illustrate its potential in more 'thick descriptive' media studies that attempt to consider 'translocal' cases within the contexts of global forces and historic flows (Kraidy and Murphy 2008).

Combining the latest research developments from cognate fields such as computational and corpus linguistics with more established empirical analyses within media studies has been a challenging methodological exercise. Whereas a certain level of subjectivity and limitation in terms of building new analytical models is inherent and inevitable in the corpus data–based approach, it remains a worthy ambition to increase the importance attached to the largely unsupervised corpus data annotation and statistical modelling in order to enhance the level of objectivity and the ability to potentially build new analytical instruments through the corpus data–driven methodological approach.

In essence, the analytical procedures deployed in the corpus data–based and the corpus data–driven approaches complement each other, and the data interpretation process can be assisted effectively by the deployment of new visualisation features furnished by popular corpus analysis tools and online platforms. Graphical analytical techniques such as mapping of data points (sets of media documents) and visualisation of the strength of association of word pairs and the strength of clusters of annotation categories have been used extensively with the corpus data–based approach. Similar visualisation techniques which are supported by exploratory statistical methods can be deployed to facilitate the pattern recognition and extraction process. The methodological advantage of visualisation techniques increases as the size of the data sets grows.

As far as visualisation in this book is concerned, we rely largely on histograms to visualise patterns in the distribution of certain annotated lexical classes, as they emerged as high- and low-frequency word groups in the data-driven analysis. The following explains in detail some of the important features at the heart of the corpus data–based and the corpus data–driven approaches to environmental media analysis.

Framing analysis, which is an essential research technique in media studies, has been integrated into the process of quantitative media analysis in order to reach a higher level of objectivity in the qualitative analysis of the empirical data. Detailed explanations are provided on how language-specific data collected from different media sources are coded and modelled for the purpose of pattern recognition and extraction. This enables future research to replicate and experiment with similar methodologies and techniques.

The variations identified across the different annotation frameworks adopted in the following chapters reflect different angles and perspectives in textual analysis. This can lead to the discovery of distinct sets of textual patterns and corpus findings in empirical text analysis. In this sense, conclusions made are largely contingent upon the structure and size of the data sets constructed and used. It would nevertheless be wrong to jump to the conclusion that, the larger the data set, the more solid or useful the corpus findings, as with the rapid expansion of online digital news and media data, predefining or setting the threshold level of a digital media database remains fraught with difficulty.

On the contrary, the practical assessment of the validity and productivity of the design of a specific research project very much depends on how effectively the analytical instruments developed are able to advance the textual analysis – in the case of corpus data–driven approach; and how the analytical tools constructed contribute to the verification or expansion of the existing theoretical frameworks – in the case of corpus data–based approach. For these reasons, we propose and endorse a pragmatic and versatile research attitude and philosophy when selecting and using either the corpus data–based or the corpus data–driven approach in environmental media analysis. That is to say, in the presence of strong and relevant theoretical frameworks, the corpus data–based approach offers the most productive analytical avenue to strengthen and enhance existing analytical frameworks; in the absence of a solid theoretical base or if the aim of the project is to identify new features that have been rarely explored before, then the corpus data–driven approach to quantitative media analysis can help. This point begins to emerge from the analysis of membership development and cross-sectoral partnership building in relation to the Great Barrier Reef in Chapter 3, and becomes fully integrated into Chapter 5's analysis of risk and responsibility discourses in relation to Japan's response to PM2.5 air pollution.

Despite the different methodological orientations between the corpus data–based and the corpus data–driven approaches, one does not exclude each other. A combination of both approaches in the design of a research project can be fully justified, depending on the research needs and questions addressed in the research process. Indeed, the flexible use of both approaches is particularly apropos to the research focus of this book as environmental innovation is still a very new research topic which is gaining increasing attention and popularity in the media.

This book contributes to the advancement of the field by developing innovative and multidisciplinary research methodologies for the study of multisectoral interaction and partnership building around environmental issues such as climate change, energy use, change of lifestyle and values informing the relationship between environment, development and sustainability. In particular, we highlight multisectoral interaction, an analytical instrument for elucidating how different societal sectors or stakeholders align with or diverge from each other on environmental issues. Indeed, the very limited amount of current research on multisectoral interaction points to a critical gap in current environmental studies, which, in turn, suggests a lack of joint action by a range of social and political sectors and actors to tackle environmental health risks and much-needed environmental social innovation, both domestically and internationally.

Another noteworthy point is our contribution in advancing the study of multisectoral interaction by developing empirical methodologies with the power to explore the building of verifiable multisectoral interaction models. An effective multisectoral interaction model plays an instrumental role in controlling public health risks, particularly as related to pollution. The book explores different types of cross-sectoral interactions, such as between government, media, business and civil society, including non-governmental organisations (NGOs), and the general public. Chapter 4 illustrates this point through an analysis of the increasing popularity in China of environmental documentaries on air pollution and associated health risks to the public which are directed and financed privately.

When examining the media politics of environmental risk and responsibility in the Chinese context in Chapter 4 we elected to employ a method that primarily involved an in-depth qualitative analysis of a single case study. This was due to the specific methodological challenges related to conducting large-scale quantitative linguistic analysis of Chinese media sources. Simply put, the Chinese authorities' efforts to monitor and censor public discourse makes data gathering – particularly online, where news articles and social media posts can appear and then rapidly disappear – fraught with difficulty. Rather than attempt to replicate the methods employed in the Australian and Japanese cases, we instead took the opportunity to engage in a more practice-focused analysis of the articulation of environmental risk and responsibility. This allowed us to examine in closer detail how multisectoral interaction and the ongoing negotiation between state and society over discourses of environmental risk and responsibility actually play out as activists attempt to navigate the political constraints placed on them by the Chinese authorities.

We also analyse initiatives taken and led by different societal sectors and actors ranging from large and small enterprises, national and subnational political authorities, national and local media, as well as actors in civil society. Environmental social innovation and engineering is emerging worldwide, and the following offers a number of typical examples of developing and growing cross-sectoral interaction models in Australia, China and Japan, within which our cross-national study on environmental media reporting unfolds:

- *A rapidly growing business-led cross-sectoral interaction model*: innovative environmental businesses – Japan (Toshiba Company's initiatives like 'Factor 10', new environmental management concept 'T-Compass');
- *An influential government-led model*: innovative environmental policies – Japan (government initiatives like 'healthy longevity society', 'societies where every woman shines' or 'womenomics', 'smart cities' or other variations of a 'smart society') and Australia (public transport and land use policies such as walkable cities and towns);
- *A media-led and popular science–supported model*: this is largely emerging in China, as illustrated by an increasing number of health programmes on the TV and radio on more active and energy-conscious living and consumption styles in a country where environmental education is traditionally very limited (taking public transport instead of using private vehicles).

As noted above, the methodological challenge for this book has been to integrate research methodologies from cognate disciplines such as applied linguistics, international relations and security, media and journalism, politics and area studies. We developed methodologies in terms of linguistic analysis, information coding and statistical modelling in order to provide an index to quantify and describe multisectoral interaction around two scenarios of environmental issues. The key procedures followed in the media data mining draw upon methodologies already developed for and tested in corpus linguistics and empirical language studies. These include a number of key points:

First, lists of the frequencies of occurrence in the running words of the data set. This was used for the creation of keyword lists to be fed into theoretically informed data coding schemes.

Second, the extraction of naturally occurring lexical collocation patterns or associated word pairs. Word collocation patterns (based on mutual information, 'MI', scores) serve as important clues to textual features of discourses developed by different societal sectors regarding environmental risks management and innovation. Different categories of word association patterns provide a basis for analysing focuses, solutions, actions and agents in the media data sets constructed.

Third, the extraction of top frequency collocates of the keywords identified important textual patterns regarding the reporting of complex pollution pathways and socio-economic patterns of environmental deterioration, as illustrated in the way top collocates (nouns, verbs, place names, adjectives) are closely related to the investigation process followed by media actors.

Fourth, the extraction and visualisation of clusters of semantic coding categories based on textual mining. Statistical methods such as exploratory factor analysis (EFA) and principal component analysis (PCA) provide an especially fruitful way for extracting clusters of semantic coding categories in different data sets from various societal sectors. The patterns of the combination of semantic coding categories enabled us to map and infer the strategies developed for different sectors to manage environmental risks and to initiate and sustain environmental innovation (Figure 1.1). The data coding schemes deployed draw upon theoretical frameworks well developed for environmental security

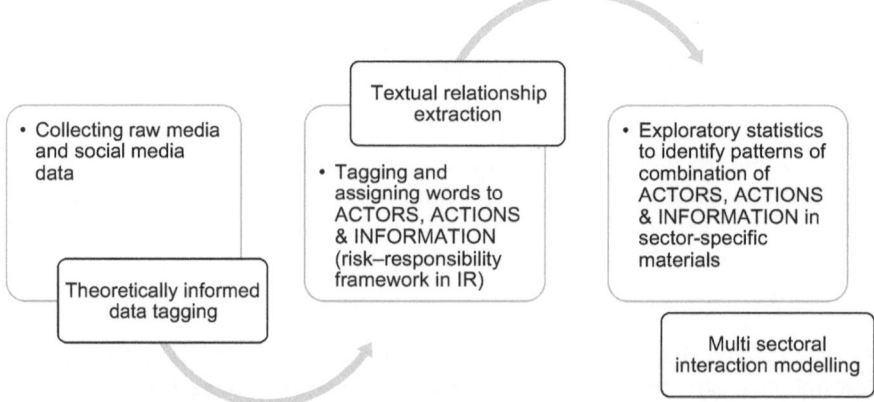

Figure 1.1 Key procedures of the empirical data analysis

Source: Authors

and international politics, which represents a major methodological innovation of this interdisciplinary book.

The extraction of multisectoral interaction models is largely based on the qualitative analysis of textual pattern uncovered in the statistical modelling of media materials. Figure 1.2 illustrates the qualitative analysis of contrastive textual

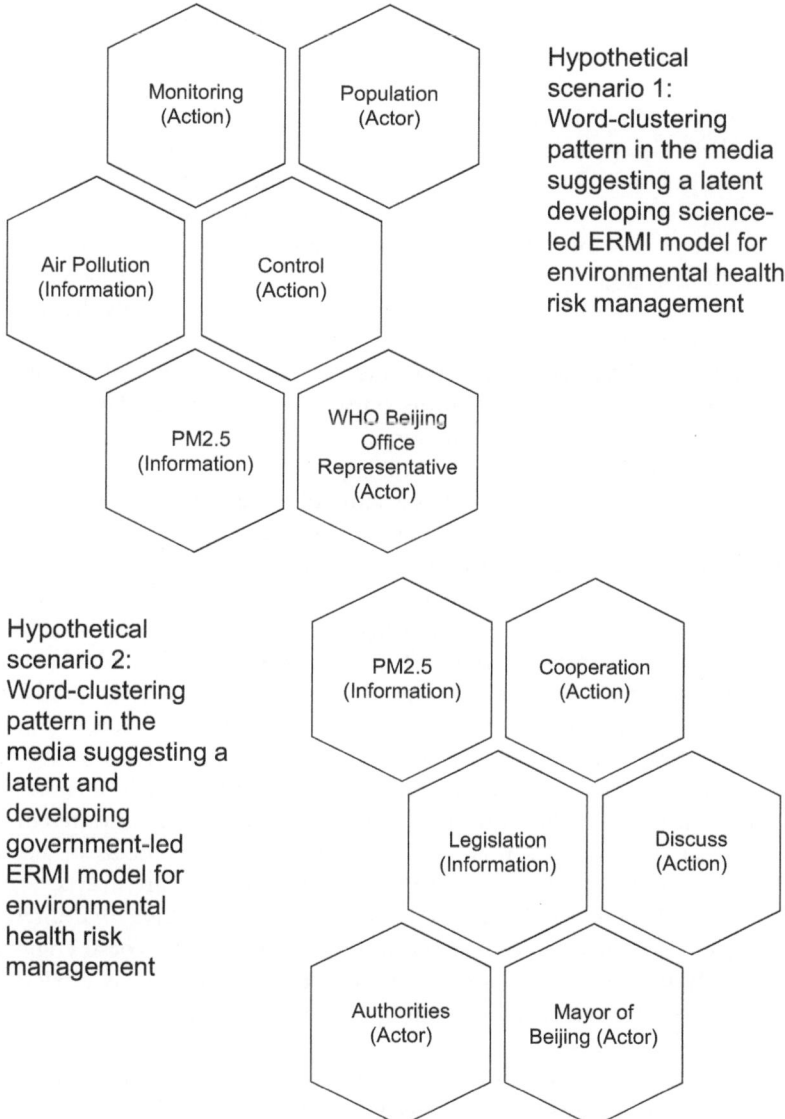

Figure 1.2 Contrastive MSI on Type 1 ERMI – framing environmental health risk in the media

Source: Authors

patterns leading to the proposition and description of distinct cross-sectoral interaction models. These models can be verified and tested with textual data from other countries and in various languages, suggesting how our work has laid the groundwork for future research in this growing research field.

A number of methodological innovations can be best introduced by reference to the work evidenced in specific chapters. Chapter 5 (Japan) represents a methodological breakthrough in discourse analyses through the use of hybrid methods to achieve multiple layers of triangulation and a conceptual model able to explain the mechanism of mediated discourse transition from risk to (self-) responsibility. It elucidates the shifting loci of risk and responsibility among the different agencies of the discourse associated with climate change adaptation policy in Japan, as well as the path and conditions of such transition, using statistical text analytics and critical discourse analysis (CDA). These methodologies are integrated within an original transdisciplinary theoretical framework drawing on social sciences, applied linguistics, as well as the two different mathematical–statistical theories of Bayesian and frequentist statistics. The process of transient assignment and reassignment of risk and responsibility to different social and political sectors and actors is then assessed and analysed, employing an innovative blend of discourse and statistical analytics to achieve the well-known benefits of triangulation.

Despite the intrinsic difficulty of building such a multifaceted methodology, the chapter succeeds in taking forward some of the authors' earlier work which has clearly demonstrated the potential of such an approach (see Chapter 5). Of singular note is the novel way this output explores at the granular level the interrelationship among the conceptual, spatio-temporal and semantic elements embedded in their data, comprising of an exhaustive set of articles from the four major Japanese newspapers published between August 1998 and March 2016 (the *Asahi*, *Mainichi*, *Nikkei* and *Yomiuri*), reporting on airborne PM2.5 and atmospheric pollution more generally. The results generated by this combination of each of their techniques breaks new ground by not only providing robust evidence for the efficacy of the approach adopted to the analysis of cross-border air pollution, but also illuminating new insights into the role of different social and political sectors and actors involved in the complex process of allocating and responding to the dynamics of the risk and responsibility matrix.

Similarly, Chapter 6 offers an exploratory study of multisectoral interaction and further partnership development around environmental social engineering in Japan. It focuses on the use of development buzzwords, as pioneered by researchers of political anthropology, political sociology, humanitarian studies and international development, to examine proffered solutions to climate change – namely, the smart society – and its impacts, which previous chapters have highlighted. While the main focus of the chapter is on Japan, it demonstrates how 'smart' solutions exist within the three geographical foci of this book – Australia, China and Japan. Further, the rich empirical data offers convincing evidence of how 'smart' may be considered a development buzzword denoting attempts to combine environmental sustainability and economic growth through the improvement of public administration and social infrastructures through information communication technology (ICT), pioneered by public and private actors, and where its exact definition and implications remains unclear and, perhaps inescapably, political.

As with research on buzzwords and development policy for developing nations, the aim of Chapter 6 is not to downplay the necessity of overcoming the problems but to draw attention to proffered solutions by decomposing the comfortable rhetoric used by different social and political sectors and actors so as to assess the feasibility, potential, corrigibility and scope of the proposed solutions in the context of multisectoral interactions. This is illustrated by the way legislative changes carried out by the Japanese state aim to utilise, and hence interact, with the private sector, to engineer market conditions so as to mobilise efforts towards a higher level of efficiency, efficacy and economic sustainability in terms of the inseparable governmental components of administration and resource. These beneficial interactions between and among the state, market and society can be expected to lead to the mutually beneficial conditions of business development and improvement of the quality of life in civil society. Significantly, the wide application of the buzzword 'smart' to initiatives, innovations or individuals provides space for political expedience with regards to the recalibration of risk associated with recent changes in Japan's public administration.

Government-led multisectoral interaction based on technological innovations to address environmental problems demonstrates important overlaps with the Japanese reporting of air pollution risks, as detailed in Chapter 5. Within the analyses presented in Chapters 5 and 6, the fundamental role of the market sector and actors in disseminating new practices and responses to environmental hazards is highlighted, where initiatives employ discourses promoting 'self-actualization' or awareness, taken to be related to notions of 'self-responsibility' indicative of the political act of transferring responsibility of the state to other sectors. In general, it seems that more possibilities and models of multisectoral interaction exist or are worth further empirical exploration with the emerging global trend of environmental social engineering and innovation, as well as state responses to major advances related to information and communication technologies.

Conclusion

While the geographical focus of this book is the Asia-Pacific, the methodological tools and instruments our interdisciplinary team deployed present a number of implications for the study of environmental discourses in different spatial and temporal zones. To start with, the findings of the research presented in the case studies in Chapters 3, 4 and 5 have been enabled by research in media, natural language processing technologies, document analysis tools and a broad range of qualitative approaches. The empirical work exemplifies the growing trend of interdisciplinary collaboration between the humanities, social sciences and information technology. The methods employed have led to the discovery of latent new social phenomena and mechanisms such as multisectoral interaction around environmental health risks, the range of risks and potential harms posed by pollution, along with rich detail of how responsibilities to deal with these risks share similarities as well as exhibit differences across our three case studies of Australia, China and Japan. Finally, in response to the challenges faced, significant environmental social innovation has been designed to grapple with both the causes and harm of pollution, as seen especially in the case of Japan. We hope our work will stimulate further theoretical, methodological and empirical endeavours to this important 'bridge-crossing' area of interdisciplinary research, as well as contribute to policymaking and civil society undertakings to tackle the risks posed by pollution in the Asia-Pacific and further afield.

References

Allan, Stuart (2002) *Media, Risk, and Science*. Philadelphia, PA: Open University Press.

Anderson, Alison (2013) *Media, Culture and the Environment*. London: Routledge.

Beck, Ulrich (2006) *Cosmopolitan Vision*. Cambridge, UK: Polity Press.

Beck, Ulrich (2009) *World at Risk*. Cambridge, UK: Polity Press.

Biber, Douglas (2012) *Corpus-based and Corpus-driven Analyses of Language Variation and Use*. Oxford, UK: Oxford University Press.

Boykoff, Maxwell T. (2013) 'Public enemy no. 1? Understanding media representations of outlier views on climate change'. *American Behavioral Scientist* 57(6): 796–817.

Boykoff, Maxwell T. and Jules M. Boykoff (2007) 'Climate change and journalistic norms: A case-study of US mass-media coverage'. *Geoforum* 38(6): 1190–1204.

Boykoff, Maxwell T. and Michael K. Goodman (2009) 'Conspicuous redemption? Reflections on the promises and perils of the "celebritization" of climate change'. *Geoforum* 40(3): 395–406.

Carvalho, Anabela (2007) 'Ideological cultures and media discourses on scientific knowledge: Re-reading news on climate change'. *Public Understanding of Science* 16(2): 223–243.

Cottle, Simon (2013) 'Environmental conflict in a global, media age: Beyond dualisms', in Libby Lester and Brett Hutchins (eds) *Environmental Conflict and the Media*. New York: Peter Lang, pp. 19–33.

Doyle, Julie (2011) *Mediating Climate Change*. Burlington, VT: Ashgate Publishing, Ltd.

Ewart, Jacqui and Hamish McLean (2015) 'Ducking for cover in the "blame game": News framing of the findings of two reports into the 2010–11 Queensland floods'. *Disasters* 39(1): 166–184.

Fraser, Nancy (2007) 'Transnationalizing the public sphere: On the legitimacy and efficacy of public opinion in a post-Westphalian world'. *Theory, Culture & Society* 24(4): 7–30.

Funtowicz, Silvio and Roger Strand (2011) 'Change and commitment: Beyond risk and responsibility'. *Journal of Risk Research* 14(8): 995–1003.

Hallin, Daniel C. and Paolo Mancini (2004) *Comparing Media Systems: Three Models of Media and Politics*. Cambridge, UK: Cambridge University Press.

Hansen, Anders (2011) 'Communication, media and environment: Towards reconnecting research on the production, content and social implications of environmental communication'. *International Communication Gazette* 73(1–2): 7–25.

Holton, Avery, Brooke Weberling, Christopher E. Clarke, and Michael J. Smith (2012) 'The blame frame: Media attribution of culpability about the MMR–autism'. *Health Communication* 27(7): 690–701.

Hulme, Mike (2009) *Why We Disagree about Climate Change: Understanding Controversy, Inaction and Opportunity.* Cambridge, UK: Cambridge University Press.

Hunston, Susan and Gill Francis (2000) *Pattern Grammar: A Corpus-driven Approach to the Lexical Grammar of English* (Vol. 4). Amsterdam: John Benjamins.

Hutchins, Brett and Libby Lester (2015) 'Theorizing the enactment of mediatized environmental conflict'. *International Communication Gazette* 77(4): 337–358.

Kay, Samuel, Bo Zhao, and Daniel Sui (2015) 'Can social media clear the air? A case study of the air pollution problem in Chinese cities'. *The Professional Geographer* 67(3): 351–363.

Kim, Sei-Hill, John P. Carvalho, and Andrew C. Davis (2010) 'Talking about poverty: News framing of who is responsible for causing and fixing the problem'. *Journalism & Mass Communication Quarterly* 87: 563–581.

Kraidy, Marwan M. and Patrick D. Murphy (2008) 'Shifting Geertz: Toward a theory of translocalism in global communication studies'. *Communication Theory* 18: 335–355.

Laviosa, Sara (2002) *Corpus-based Translation Studies: Theory, Findings, Applications.* Amsterdam and New York: Rodopi.

Leiserowitz, Anthony A., Edward W. Maibach, Connie Roser-Renouf, Nicholas Smith, and Erica Dawson (2013) 'Climategate, public opinion, and the loss of trust'. *American Behavoural Scientist* 57(6): 818–837.

Lester, Libby (2015) 'Three challenges for environmental communication research'. *Environmental Communication* 9(3): 392–397.

Lester, Libby (2016) 'Containing spectacle in the transnational public sphere'. *Environmental Communication* 10(6): 791–802.

Lester, Libby and Simon Cottle (2009) 'Visualising climate change: TV news and ecological citizenship'. *International Journal of Communication* 3: 920–936.

Liang, Xuan, Jiun-Yi Tsai, Kristine Mattis, Magda Konieczna, and Sharon Dunwoody (2014) 'Exploring attribution of responsibility in a cross-national study of TV news coverage of the 2009 United Nations climate change conference in Copenhagen'. *Journal of Broadcasting & Electronic Media* 58(2): 253–271.

Nanz, Patrizia and Jens Steffek (2004) 'Global governance, participation and the public sphere'. *Government and Opposition* 39(2): 314–335.

Nisbet, Matthew C. (2009) 'Communicating climate change: Why frames matter for public engagement'. *Environment: Science and Policy for Sustainable Development* 51(2): 12–23.

Olausson, Ulrika (2009) 'Global warming-global responsibility: Media frames of collection action and scientific certainty'. *Public Understanding of Science* 18: 421–436.

Olausson, Ulrika (2011) '"We're the ones to blame": Citizens' representations of climate change and the role of the media'. *Environmental Communication* 5(3): 281–299.

O'Neill, Saffron and Sophie Nicholson-Cole (2009) '"Fear won't do it" promoting positive engagement with climate change through visual and iconic representations'. *Science Communication* 30(3): 355–379.

Painter, James (2013) *Climate Change in the Media: Reporting Risk and Uncertainty.* Oxford, UK: Reuters Institute for the Study of Journalism.

Pantti, Mervi, Karin Wahl-Jorgensen, and Simon Cottle (2012) *Disasters and the Media.* New York: Peter Lang.

Pollack, Henry N. (2005) *Uncertain Science . . . Uncertain World*. Cambridge, UK: Cambridge University Press.

Schmidt, Andreas, Ana Ivanova, and Mike S. Schäfer (2013) 'Media attention for climate change around the world: A comparative analysis of newspaper coverage in 27 countries'. *Global Environmental Change* 23(5): 1233–1248.

Schmidt, Andreas and Mike S. Schäfer (2015) 'Constructions of climate justice in German, Indian and US media'. *Climatic Change* 133(3): 535–549.

Starke, Christopher and Felix Flemming (2015) 'Who is responsible for doping in sports? The attribution of responsibility in the German print media'. *Communication & Sport*. Published online September 3.

Tranter, Bruce and Libby Lester (2015) 'Climate patriots? Concern over climate change and other environmental issues in Australia'. *Public Understanding of Science*. Published online December 15.

Waisbord, Silvio (2016) 'Translations: Communication studies without frontiers? Translation and cosmopolitanism across academic cultures'. *International Journal of Communication* 10: 19.

Yu, Hongsik, Jingren Si, and Jaehee Cho (2015) 'Effects of emotional exemplars in responsibility attribution-framed news reports on perception and evaluations of social issues'. *Asian Journal of Communication* 25(5): 525–545.

Zhang, Yuan, Yan Jin, and Yunbing Tang (2015) 'Framing depression: Cultural and organizational influences on coverage of a public health threat and attribution of responsibilities in Chinese news media, 2000–2012'. *Journalism & Mass Communication Quarterly* 92: 99–120.

2 Risk, responsibility and pollution in Australia, China and Japan

A regional overview

This chapter provides an overview on how the three countries at the core of this study – Australia, China and Japan – have faced and addressed a range of environmental issues and the risks posed to society by pollution and other manifestations of environmental degradation, including climate change. It seeks not only to identify how these issues have emerged within the domestic context, but also to clarify the locus of responsibility in tackling pollution and other environmental risks, as illustrated by the actions taken by a range of different stakeholders, including governments, media, non-governmental organisations (NGOs) and local citizens. In this way, the chapter sets the scene for the case studies to follow.

We start with a discussion of Australia. The environmental risks faced by the Australian government and people are largely a result of the environmental challenges resulting from the exploitation of natural resources for not only Australia's own growth, but also to support economic development elsewhere, including in China and Japan. The modern environmental movement in Australia started in the 1960s, and sought to protect landscapes from development and stop the export of natural resources such as uranium. In the 1980s the environmental movement protested the risks and potential harms from the construction of dams, such as the successful action to stop the damming of the Franklin River in Tasmania, which brought the Australian environmental movement to international attention. The mediated environmental politics to grow out of such action have exerted a long-term influence on the responsibility to address environmental risks up to this day. Now, environmental risks posed to the Great Barrier Reef are of central concern to not only environmental activists, but the central and state governments, as well as the media and people more broadly. In this context, the media have reported the national division in opinion on how to protect the reef from the risks posed, illustrating the wider ongoing challenge to Australia as a resource-based economy, which faces the task of striking a balance between access to resources (minerals, gas, coal and timber) and the responsibility to take action to tackle climate change at the national and global levels.

The section on China takes a different focus. China's modernisation during the post-1949 years has recently seen rapid economic growth, but development in the last two or three decades has been accompanied by the emergence of a range of environmental risks, especially man-made risks related to atmospheric pollution. In addition, risks posed by flooding and other natural disasters have been faced for millennia, with dam building, desertification and deforestation creating different kinds of environmental risks and harms for the country in the recent past. These environmental risks, potential harms and impacts have been addressed over time by a range of actors, not only the government, but also NGOs and so-called government-organised non-governmental organisations (GONGOs) as well as ordinary citizens. These actions at the non-governmental level have been stimulated by the diversification of information now available to the citizens through the Chinese media, which have become a much more important source of critical data on the environment than in the past.

The final section of this chapter focuses on Japan. It reviews how the path of modernisation pursued by the government during the late nineteenth century led to a range of environmental risks and harms to the population, as illustrated by the impact on people's health caused by industrial waste from copper mining polluting the waterways. The protest movement to emerge in opposition to the risks posed by the Ashio mine gradually moved from a focus on eliminating the pollution to making the mining company responsible for compensating the people for the harm caused. Such popular protests going back to the late nineteenth century are the background for the numerous post-war environmental movements, with the 'big four' pollution cases coming to public attention in the 1960s. The media served to galvanise public opposition to the costs to the population of the growth strategy pursued by the government. More recently, the media have been at the centre of the controversy over the risks and harms produced by the Great East Japan earthquake, tsunami and nuclear crisis of 2011 onwards. While this has boosted the supply of coal from Australia, the protests against the restarting of nuclear power plants has helped to change public support for Japan's reliance on nuclear energy.

Australia

Australian history is dominated by stories of hardship imposed by landscape. Drought, fires, floods and cyclones are the 'natural' risks that accompany living on a dry, island continent swept by Antarctic lows and the Roaring Forties in the south, long droughts followed by destructive floods brought to the inland by the El Niño–La Niña oscillation and oppressive wet seasons in the tropical north. Australia is the world's sixth-largest country in terms of land size – 7.6 million square kilometres in comparison to China's 9.4 million square kilometres and Japan's less than 400,000 square kilometres – but its population remains relatively small at just over twenty-three million (in 2013). Tensions between population growth and the capacity of the land to sustain the cities that largely hug the continent's east coast (Melbourne, Sydney and Brisbane) have been continuing. The trials for farmers that come with being 'on the land' are central to the Australian psyche, evident in the country's best-known literature, paintings and other artworks, with the landscape and weather often depicted as threatening, alien and unpredictable. The land could bring great rewards, but only to those resilient enough to overcome the many challenges. Wool growing was European Australia's first success in 'beating' the land – so much so that the country is often described as having been built on the 'sheep's back' – with mineral resources, including gold and coal, quickly becoming a major export earner. Iron ore remains the county's most valuable export, followed by coal and gold.

From the earliest years of British colonisation in the late eighteenth century, however, concern about human-induced harm to the environment and accompanying risks to its inhabitants has also been part of the Australian media, if less so part of the national narrative. The removal of native trees or ferns from landscapes considered aesthetically pleasing to settlers emerged as an issue, along with overhunting of economically valuable whales and seals. Dingoes, thylacines (Tasmanian tigers) and eagles, in contrast, were considered pests, undermining the emerging wool industry by picking off the flocks' most vulnerable. By the mid-nineteenth century, environmental press coverage was dominated by stories of urban pollution and contamination of the too-few vital waterways. Individual brewers, tanners and later industrialists were commonly named as responsible (Lester 2014).

In the late nineteenth and early twentieth centuries, a love of nature for nature's sake also found a place in the Australian press. Bird watchers, flower painters and mountain climbers were given a voice to describe the Australian landscape to the increasing number of Australians isolated from the natural environment in the major cities (Bonyhady 2003; Meadows and Thomson 2013). A sense of melancholy, nostalgia and loss pervaded these stories – just as they pervaded writings about indigenous Australians, who at that time were still experiencing a systematic programme of forced removal and alienation from their traditional lands. That this was occurring after tens of thousands of years of sustained environmental management by Australia's indigenous peoples was an irony lost in these writings.

The contemporary environment movement that emerged in Australia in the 1960s was inspired in part by the anti-nuclear and air pollution anxieties of the European movement, and in part by the wilderness campaigns of North America. Australia's Westminster-style political system, alongside its structure as a federation of states and territories, provided opportunities for candidates focused on a single issue to gain seats in parliament – via the houses of parliament at a federal level and in most states that are elected through proportional representation. The United Tasmania Group, with its detailed environmental manifesto, sought to influence political outcomes by standing candidates in elections in the early 1970s in the southern island state of Tasmania, and is now recognised as the world's first green party (Lohrey 2002).

Throughout the 1960s and 1970s, bitter local and national campaigns were fought to protect landscapes from development over sand mining on Fraser Island at the southern end of the Great Barrier Reef, the Daintree Rainforest on Cape York in the far north, wet eucalypt forests in New South Wales and the damming of Lake Pedder in south-western Tasmania for hydroelectricity. These battles were fought alongside campaigns to prevent uranium exports and to protect early colonial urban precincts in Sydney. Known as 'green bans', they were imposed as a form of environmental activism by the union movement that, at the time, wielded substantial power over Australia's building industry, and were the first of their type in the world. Nevertheless, the alliance of the labour and green movements that formed for the action has rarely been repeated, with tensions between the jobs that accompany development and conservation remaining unresolved.

Within months of the *New York Times* appointing the world's first 'environment' reporter, the Melbourne *Age* had appointed Australia's first dedicated reporter to cover the contentious issues that were increasingly together under the umbrella term 'the environment'. These stories were framed as political, economic and science-based: who owned the land and its riches; who had the right to exploit or conserve it and who made policy to govern it and on the basis of what information. The fact that all three capabilities often sat with a small number of elite landowners and politicians meant that Australian environmental reporting tended to focus on abuses of power and corruption, and was often accused of being 'anti-development' (Lester 2013). The conflicted relationship between environmental coverage, Australia's often conservative press and industry and government was established during this period, and continues to influence media reporting of environmental issues in mainstream Australian media.

The Franklin River campaign of the early 1980s cemented the environment within Australia's political and media agenda as contentious, strategic and highly political. The well-organised campaign was fought over several years to prevent the damming of the remote south-western Tasmanian river, and saw the fall of two Tasmanian state governments while also playing a prominent role in national politics. Protests culminated over the southern summer of 1982–1983 in a physical blockade of the construction site. More than 1,200 people were arrested 'in the wilderness' and carted by boats, buses and paddy wagons to distant courts and gaols for processing, while coverage ran on front pages and led news bulletins around the country. International celebrities, such as Spike Milligan and David Bellamy, joined the campaign, ensuring wide attention for the cause. Indeed, the environmental campaign has been described as the first in the world to attain 'global stature' (Hay 1991). The United Nations Educational, Scientific and Cultural Organization (UNESCO) inscribed the area on its World Heritage list in 1982, providing a newly elected federal Labor government the excuse to intervene formally in the Tasmanian dam-building proposal. The debate inevitably turned into one of states' rights versus federal interference. The federal government won the bitter battle in June 1983, with a High Court of Australia decision that upheld the nation's international responsibilities. Construction was stopped.

The Franklin case established a pattern for mediated environmental politics that exists in Australia today (Lester 2007). This includes the right and responsibility of local and state governance structures to measure levels of concern and determine degrees of risk in relation to social and economic return; the often conflicting interaction of local and state environmental law, policy and politics with national interests; the ambition of campaigners, industry and governments to gain or contain international interest and interference in Australian conflicts; and the important role of news and other media as both an arena for public debate but also as a political actor, pushing for action, circulating or suppressing powerful symbols and meanings, investigating or ignoring environmental breaches, establishing and undermining hierarchies of expert credibility, and attributing or denying blame.

During the Franklin debate, for example, Tasmania's major metropolitan newspaper, the Hobart *Mercury*, campaigned heavily in support of the dam and the rights of the state to control the uses of land and resources. In almost daily editorials, it urged Tasmanians to vote against the federal Labor Party, which was campaigning on a platform to stop the dam if elected, while also undermining through ridicule the mainland and international politicians, celebrities and protesters who were flocking to Tasmania to oppose the dam's construction. The protests were described as a meaningless, staged performance, enacted only 'for the cameras'. In its news reporting, the Tasmanian newspaper avoided repeating terms and other discourse strategies favoured by the environmental movement, in particular its labelling of the threatened area as 'wilderness'. Meanwhile, the Melbourne *Age*, at the time Australia's most influential paper among the educated, liberal voters of mainland Australia, campaigned against the dam, claiming the right of its readership to have a voice in the region's future. It repeatedly adopted the environmental movement's discourse strategy of describing the area as 'wilderness', shifting the term from its previous connotations with barren uninhabitable places to something desirable and valuable for its own sake (Lester 2005). The World Heritage area, which covers more than 20 per cent of Tasmania's landmass (a further 10 per cent is reserved in national parks outside the World Heritage area) and known as the Tasmanian Wilderness, includes more UNESCO values than any other inscribed site, and the somewhat paradoxical industry of 'wilderness tourism' became one of the island's most valuable sources of income soon after the High Court decision that stopped the dam.

Australia's environmental politics continues to reflect many of the features that emerged during the campaign over the Franklin dam. The health and management of the Great Barrier Reef, the world's largest and most spectacular reef, stretching along the Queensland coast, has been prominent within recent environmental debate (see Chapter 3). Other issues include commercial access to water as the country's agricultural industry becomes increasingly irrigation dependent. The intersection between indigenous and environmental issues in Australia is complex, particularly given increasing pressure for resource and agricultural access. Aboriginal communities, for example, have combined with environmentalists to fight the risks posed by the construction of a gas plant in the Kimberley region of far north Western Australia, while the conservation of wild rivers on Cape York peninsula in the far north has been legally challenged on the grounds that it violated indigenous community rights to develop the rivers and sell water rights.

Access to resources, such as minerals, gas and timber, is set to continue as a major source of political conflict in Australia. Indeed, given Australia's desire to capitalise on the burgeoning middle classes of Asia, such conflict is expected to intensify. A 2012 government white paper, while optimistic about the national prosperity the 'Asian Century' would bring, acknowledged that a rising trade of raw materials, manufactured goods, ideas and people faced some challenges and risks, among them regional conflict as Asian countries competed for limited resources, including water and minerals; increased pressure on Australia's resources and infrastructure; and environmental degradation that could hinder the country's capacity to meet demand (Australian Government 2012; Lester 2014). Japan, in particular, has sought to increase access and security of supply to Australian coal since the Fukushima disaster in 2011 that shut down the country's nuclear power plants. Australia responded to such pressures by, in the words of then Prime Minister Tony Abbott in his 2013 victory speech, being 'open for business'. Meanwhile, the *Guardian*'s Australian edition produced 'an activist map of Australia, charting environmental protests going on around the country right now. If you know about a protest near you, please tell us' (*Guardian* 2014).

Biodiversity and habitat loss, vulnerability to natural disasters, especially cyclones, droughts and bushfires, and population growth are all debated within the context of climate change impact. The Great Barrier Reef has acted as a powerful symbol in the extreme politics that has played out in Australia over climate change. Before becoming prime minister in 2007, Kevin Rudd suggested that climate change was the 'greatest moral challenge of our time'. Yet, his short tenure as prime minister coincided with the most recent resources boom and an unprecedented level of mining activity centred on coal. Australia largely avoided the effects of the global financial crisis as a result. Climate change mitigation measures have now played a significant role in the downfall of two prime ministers, including Rudd, and a leader of the opposition. Under Rudd's successor, the conservative Tony Abbott, the Australian government worked internationally to promote a 'coalition of the unwilling' on carbon minimisation, and imposed a modest reduction target of 5 per cent by 2020 despite Australians being among the biggest emitters of carbon per capita in the world.

However, as Australia's media contain a notably high presence of climate change sceptics and deniers, debate often focuses on the veracity of science and economic modelling rather than mitigation measures (McKnight 2010; McGaurr *et al.* 2013). Contemporary Australian media's coverage of environmental issues has always been marked by a form of reluctant symbiosis. The mix of risk, conflict, impact, proximity, science and politics embedded in many environmental stories is a heady one for news media, and made even more desirable by the imagery and symbolism of palm trees bent by cyclonic winds, stock animals wasted by drought-caused famines and office workers wading through flooded streets.

Environmental groups learned the lessons of early campaigns and developed a repertoire of actions and images that could be captured by media and circulated. The possibilities of the Internet were quickly incorporated into this repertoire, delivering visuals and social media statistics to further enhance the news value of environmental stories (Lester and Hutchins 2009). Nevertheless, Australia's resource-based economy has ensured that government and industry maintain intense scrutiny of environmental coverage, with 'flak' – in the Herman and Chomsky (2010) sense of elites maintaining control – commonly reported by environmental journalists in Australia, alongside a form of self-censorship within newsrooms (Lester 2013).

This situation is not helped by the fact that the Australian news media market is one of the world's most concentrated, with newspaper and associated digital markets dominated by two companies, Rupert Murdoch's News Corp and Fairfax Media Limited. In the politically influential national and metropolitan markets, Murdoch titles comprised 65 per cent of circulation in 2011 and Fairfax 25 per cent (Flew and Goldsmith 2013). Print's share of advertising has dropped since 2007 when it controlled half of all advertising expenditure in Australia (Australian Communications and Media Authority n.d.). Commercial television and radio in Australia is protected from market cornering to some extent by cross-media ownership laws that prevent companies from owning more than one or two types of media assets (such as radio and television licences) in the same area. However, new digital players have shifted the market considerably. By 2011, for instance, revenue for the six major operators of free-to-air television revenue had declined A\$1.2 billion from a peak of A\$5.77 billion in 2005–2006 (Australian Communications and Media Authority n.d.), despite Australians watching on average three hours of television a day (free-to-air and/or paid).

News and current affairs delivered by the national public broadcaster, the Australian Broadcasting Corporation (ABC), is always under intense scrutiny by the conservative side of politics. Climate change coverage has been a specific focus of critics. Likewise, the country's only general news national newspaper, *The Australian*, owned by Murdoch's News Corp, is criticised for its reporting of climate issues (McKnight 2010). Overall, a comparative study of reporting of climate risk and uncertainty undertaken by Oxford University identified Australian media as carrying a higher proportion of climate sceptics and deniers than other countries included in the survey (McGaurr *et al.* 2013).

Environmental journalism in Australian news media has suffered the decline also seen in many other national markets, with the dedicated role disappearing from most newsrooms. Only the ABC and Fairfax Media and News Corp flagship titles (*Australian*, *Age* and *Sydney Morning Herald*) continued the round post-2011 (Lester 2013). The failure of Conference of the Parties 15 (COP15) in Copenhagen, changes in federal political leadership and an interruption to El Niño–related drought saw interest wane. Indeed, the Australian federal election of 2013 was described as the 'election that forgot the environment' with a notable decline from 2010 in media coverage of environmental stories, including those focused on climate change (Lester *et al.* 2015). Despite significant activity by environmental groups behind the scenes, this did not translate into a presence in news media during the 2013 campaign coverage, with environmental movement organisations appearing less in media coverage in the lead up to the election than they had in any election since 1990. The issue receiving by far the most coverage was climate change, followed by a raft of issues that included World Heritage, Arctic drilling, coal seam and shale gas exploration, clean water and air, pollution, recycling, forestry and the Great Barrier Reef.

Public opinion on environmental issues in Australia has, to some extent, followed international trends, with noted peaks of concern in the late 1980s and mid-2000s. Results from the 2013 Australian Survey of Social Attitudes suggest that Australians prioritise climate change and pollution as their most urgent environmental issues, with the level of climate change concern highly circumscribed in terms of the social and political backgrounds of its supporters/detractors. Women are more likely to choose climate change as their most urgent environmental issue, as are the tertiary educated. Overpopulation, marine conservation, extreme weather events, destruction of wildlife, waste disposal, mining, logging, soil degradation, nuclear power and loss of biodiversity were other issues identified as 'most urgent' (Tranter and Lester 2015).

The source of information about environmental issues clearly impacts upon issue priority in Australia – those who nominated scientists or environmental organisations as their most trustworthy source were more likely to prioritise climate change. Interestingly, survey results also showed that although there have been recent attempts in Australia – as in the United States and elsewhere – to link action on climate change with a love of one's country, a correlation exists in Australia between patriotism and denial of climate change as a significant issue. Indeed, expressing concern over climate change appears to be unpatriotic for some Australians. Even after controlling for political party identification and other important correlates of environmental issue concerns, survey results find that patriots are less likely than others to prioritise climate change as their most urgent environmental issue and less likely to believe that climate change is actually occurring (Tranter and Lester 2015).

China

As one of the largest countries in the world, covering approximately 9.6 million square kilometres, China's natural environment includes great geographical and biological diversity. It contains major deserts such as the Gobi and Taklamakan as well as the Tibetan Plateau and Himalayas, the grasslands of Inner Mongolia, the tropical beaches of Hainan and the dense forests of Sichuan, home to the iconic giant panda. The vast majority of the Chinese population lives in the eastern part of the country, yet even here there is significant variation in the landscape. In the dry north, wheat and corn are staples and the winters are brutally cold, while in the warm south the abundance of water has led to the development of a sophisticated system of rice cultivation. The north and south are divided by the Yangtze River, which links China's mountainous interior to the major cities of Chongqing, Nanjing and Shanghai. China's other major river, the Yellow River, lies in the north of the country, and its fertile surrounds are considered the cradle of Chinese civilisation.

China has a long history of dealing with environmental risks. Major disasters, particularly floods, have periodically claimed large numbers of lives. Emperors, in conducting rituals to maintain harmony between humanity and nature, represented the locus of responsibility for addressing environmental issues, and when natural disasters struck they could be viewed by the people as a sign that the ruler had lost the 'mandate of heaven' that was the source of his legitimacy. The silty Yellow River is especially prone to flooding and has traditionally been referred to as 'China's sorrow' due to the disasters that have befallen those who live near its banks. Dams and other man-made means of controlling river flows have helped reduce flooding but at times have created new vulnerabilities. In an infamous incident in 1938 the Nationalist army, in an attempt to halt the advance of the invading Japanese forces, deliberately destroyed dykes on the Yellow River, causing floods that resulted in the deaths of tens of thousands of civilians and the displacement of hundreds of thousands of people (Crossley 2010: 199; Muscolino 2015: 2).

Attitudes towards the environment in China have been influenced by the interaction between an eclectic mix of local and imported philosophical traditions. The Daoist philosophy, which emphasises harmony with nature, was a traditional counterpoint to the dominant traditions of Confucianism and Legalism, which take a more utilitarian and anthropocentric approach to the natural world. The combination of socialism and nationalism that became the primary ideological force as China emerged from civil war in the mid-twentieth century had profound consequences for the natural environment. Whereas in the nineteenth century the nationalist cause of protecting the country from foreign aggression had sometimes been linked with resistance to modernisation, as when officials objected to the building of railways due to fears of invading troops exploiting their use to make rapid advances (Spence 1990: 250), in post-1949 China strengthening the nation became synonymous with modernisation of the economy. China's leaders prioritised economic growth in an attempt to catch up with the developed world as quickly as possible. Mao Zedong's campaigns of mass mobilisation were based on a belief that China's common people, through collective endeavour and sheer force of will, could shape nature to meet their needs and achieve rapid advancements in growth and productivity (Shapiro 2001). In the countryside this manifested in the form of mass campaigns to kill pests that had significant negative effects on local ecosystems as well as an explosion of small-scale steel production that led to large-scale deforestation to fuel the furnaces. As far as pollution was concerned, the prevailing attitude in the People's Republic of China throughout most of the twentieth century was 'pollute first, clean up later,' with the responsibility for that clean-up being left for the future.

Although environmental protection took a back seat to modernisation during the twentieth century, at times dissenters emerged to question the environmental cost of this rush to develop the economy. Some major development projects have generated political controversy due to the environmental risks posed, with the Three Gorges Dam being an early case that attracted nationwide attention (see Dai 1994). The dam on the Yangtze River was to be the largest in the world and was intended to help meet the rapidly growing need for electricity and better control flooding in the region. This dam project was the first – and so far the only – issue that has led to significant open political division in China's rubber-stamp legislature, with a third of the National People's Congress delegates either voting against the proposal or abstaining (Baum 1994: 349). True, it was not only the environmental risk of the dam but also concerns over the social cost of relocating the more than one million residents whose homes would be inundated, safety issues, corruption and the loss of local heritage sites that generated opposition to the construction among delegates and activists. Despite this opposition and a lack of support from key international organisations such as the World Bank (Jackson and Sleigh 2001: 60) the dam was eventually completed, but the controversy surrounding the project highlighted the growing political salience of the environmental and social problems associated with China's breakneck drive for modernisation.

The risks associated with these large-scale projects have not deterred the Chinese government from further attempts to reshape China's physical environment. The South-North Water Transfer Project, an ongoing construction project designed to shift around 45 billion cubic metres of water per year from China's wet southern regions to the dry north of the country (Moore 2014), is perhaps the most ambitious of the Chinese state's efforts to transform China's natural environment to meet its people's needs but it is not the only example of this kind. The Chinese state also takes responsibility for running an advanced weather modification programme that primarily aims to increase rainfall in drought-prone areas but also has a number of other functions, including reducing fog around airports and keeping the skies clear for major political and sporting events (Edney and Symons 2014: 320).

Increasing openness within China in the 1980s and 1990s created new domestic political space for the expression of ideas about environmental protection, while China's engagement with the outside world, beginning with its readmission to the United Nations in 1971 and accelerating in the post-Mao period, provided opportunities for China to interact with the global environmental movement. This is seen, for instance, in China's decision to send representatives to major international environmental forums such as the 1992 Rio Earth Summit. What is more, a wider sense of responsibility for the environment became clear when the government signed up to important global agreements, including the Montreal and Kyoto protocols. The Chinese government began to use the term 'sustainable development' in the 1990s (Edmonds 2011: 16) and has now moved away from the earlier view that environmental risks are a symptom of capitalism and therefore not present in socialist countries. It has now moved towards an acknowledgement that such risks, especially pollution, remain a major barrier to China's development. At the same time, however, the government has also strongly opposed any attempt to shift responsibility for environmental protection from developed countries to the developing world, most prominently through its position in global climate change negotiations (Edney and Symons 2014: 322).

China's long history of environmental degradation (see Elvin 2008), huge population and rapid drive for modernisation have combined to create a contemporary environmental crisis in many parts of the country. Flooding, desertification, water scarcity and deforestation are some of the most serious environmental risks faced by China today (Economy 2010: 9–10). The problem that is most visible to Chinese residents, however, is atmospheric pollution. A 2015 World Bank report found that 99 per cent of the country's population lives in areas where the mean annual concentration of PM2.5 – the particulate matter in polluted air that can penetrate deep into people's lungs and cause serious health problems (see Chapter 5 for a full discussion) – exceeds World Health Organization (WHO) guidelines (World Bank 2015: 58), and one study linked this form of pollution to 1.23 million premature deaths in China in 2010 (Boyd 2014). Another study by the Ministry of Environmental Protection estimated the direct economic losses due to pollution in 2010 at 3.5 per cent of gross domestic product (GDP), not including the indirect costs incurred due to harm to health, and found that the cost of pollution was rising faster than the country's overall GDP growth rate (Li 2013). Water pollution is also a serious concern; a government study found that approximately 40 per cent of China's freshwater remains a public health risk (cited in Zhang and Barr 2013: 4–5). Cancer has increased to become China's number one cause of death, and this increase has been accompanied by the emergence of the phenomenon of 'cancer villages', where abnormally high rates of cancer have been identified in populations living next to chemical factories (Lora-Wainwright 2010). Climate change has also become a major international issue for China. The country's carbon dioxide emissions are approximately double those of the United States, making it the world's largest aggregate emitter of greenhouse gases, and China now has higher per capita emissions than the European Union (McGrath 2014). By 2030 China is expected to account for half of all global emissions (Gilley 2012: 289).

China's integration into the global production and consumption chain has had a major impact on the environment. China's role as the world's factory has meant that a certain amount of the risks posed by pollution, such as the harmful by-products of recycling or waste from manufacturing, has been 'outsourced' from the developed world to China (Minter 2013). China has long been a key destination for trafficking in exotic and endangered species, and international conservationists have voiced concerns that as Chinese incomes rise this will lead to a corresponding increase in global demand for traditional delicacies such as shark fin and products made from animals such as rhinos and tigers. However, highly visible advertising campaigns urging Chinese to avoid consuming these products have now emerged in China and feature prominent Chinese celebrities such as the basketball player Yao Ming. In September 2015, China and the

United States signed an agreement to implement a nearly complete ban on the import and export of ivory (Ryan 2015).

As the Chinese Communist Party (CCP) has shifted away from relying on revolutionary ideology to boost its legitimacy and towards relying on improving the lives of its people it has been forced to take environmental problems seriously, although economic development still remains the more important goal. The prominent impact of pollution on people's daily lives has meant that Chinese public awareness of and concern over environmental issues is now very strong, and these issues can be a key catalyst for discontent. For instance, a 2009 poll found that, when asked to identify the major security risks facing China in the next ten years from a list of possible options, those surveyed were most likely to select 'environmental issues like climate change', with 'water and food shortages' the second most commonly identified risk (Hanson and Shearer 2009: 3). The authorities responded by taking steps to introduce the language of environmental protection into their official policy discourse, developing new legal and regulatory mechanisms designed to deal with these problems and upgrading the status of the state environmental agencies. The point is illustrated by the CCP's Seventeenth National Party Congress in 2007 when the then-General Secretary of the Party, Hu Jintao, introduced the phrase 'ecological civilisation' (*shengtai wenming*) into the Party's official model for economic development (*China Daily* 2007). In the following year the State Environmental Protection Administration (SEPA), which had gradually seen its status increase since its earliest inception as the State Council Leading Group Office for Environmental Protection in 1978, became the Ministry of Environmental Protection (Yang 2008) and the policy group tasked with coordinating the government's response to climate change has also increased in status over time (Held *et al.* 2011: 23–24).

Although an attempt by SEPA to measure 'green GDP' was eventually abandoned due to resistance from local officials and the unwillingness of the National Bureau of Statistics to provide access to accurate data (Li and Lang 2010), environmental protection measures have been introduced into the metrics used in the performance appraisals used to assess officials' work as well as their suitability for promotion. In 2015, the State Council announced that the weighting of environmental outcomes in cadre appraisals would be 'significantly increased' and that officials would be held accountable for environmental harm even after they had left their post (Li 2015).

China's system of environmental law is highly complex, and a number of different institutions have contributed to the proliferation of laws and regulations, including the Ministry of Environmental Protection, the National Development and Reform Commission and the Ministries of Water Resources and Agriculture. Despite the increase in the quantity of legislation in this area the quality of environmental laws and regulations remains low, with many of them reading more like policy statements than robust legal guidelines due to their lack of precision (Chen 2009: 38–39). China's political system of 'fragmented authoritarianism' (Lieberthal 1992) can also lead to weak enforcement. The enforcement of new laws designed to punish polluters and mandating stricter standards for industry relies on the cooperation of local officials, who often have a stronger incentive to encourage economic activity than to protect the population from environmental risks and sometimes even have a personal financial stake in polluting businesses (Economy 2007). In 2013, the minister of environmental protection, Zhou Shengxian, made the startling claim that his ministry was one of the most embarrassing government departments in the world, but at the same time he pointed out that the Ministry of Environmental Protection did not have sole responsibility for addressing many environmental risks, such as water monitoring and land use (Reuters 2013).

Non-state actors also play an important role in the politics of the environment in China. The emergence of new civil society actors began in China's reform era and accelerated in the 1990s but did not follow the same pattern as it had in Western countries or in Eastern Europe (Ho and Edmonds 2007). Many of these new actors were in fact GONGOs deeply embedded in the power structure of the party-state. Rather than act as an independent political force that could directly challenge the state, many NGOs attempt to walk a fine line between cooperation and confrontation in order to maximise their influence and minimise the possibility of being shut down by the authorities, leading to a form of 'negotiated symbiosis' between state and society (Ho and Edmonds 2007). In the late 1990s, the government acknowledged that there were many areas where 'social organisations' could play an important role by filling a gap in state provision and decided to allow the growth of the sector while putting in place stringent registration requirements to ensure that these groups could not grow into an independent political force (Ho 2001). Some groups choose not to register with the authorities, calculating that the greater freedom this provides will outweigh the risk of prosecution and hoping the positive impact of their activities on the community will increase the likelihood that local officials will tolerate their existence. Although the space for activism in China is restricted, the political system is still conducive to voluntary collective action on many social causes, including those related to environmental risks and potential harms (Ho 2007).

Growth in public awareness of environmental risks and an increase in citizen-led environmental activism have been facilitated by changes to the Chinese media that have occurred in the reform era. From the late 1970s to the 1990s the Chinese media environment went through a dramatic transformation as policies were brought in that reintroduced advertising, reduced or eliminated many of the subsidies that were propping up the state-run media and allowed commercial media outlets to emerge. During this period China's media landscape shifted from one in which there were a very limited number of media sources that all were state run to a highly diverse environment. In the 1980s an average of one new newspaper was founded every three days (Zhao 1998: 57); now there are nearly 2,000 newspapers, 10,000 magazines and more than 3,000 television stations (Zhan 2011: 116). Most of these media outlets operate largely along commercial lines rather than act as a mouthpiece for the Party, although the CCP still views the media – especially the news media – as responsible for transmitting the 'correct' information and viewpoints to the public rather than as an independent check on power.

Some Chinese media outlets are state owned and operated and others are not, but all are subject to official control through China's media licencing regime and must submit to instructions from their supervising branch of the CCP's Propaganda Department (see Brady 2008). Although most news reports are not typically reviewed by censors before broadcast or publication, there is strong incentive to self-censor due to the potential for serious punishments to be imposed on media companies as well as on individual journalists and editors. The system is characterised by grey areas and ambiguity rather than clear rules and guidelines for reporting. Although experienced journalists and editors generally have a good sense of when to push the boundaries on a particular issue and when to take a more conservative line, they often still face significant risks when reporting on controversial topics or engaging in investigative journalism.

The advent of mobile phones, the Internet and social media has made it much more difficult for the authorities to suppress public awareness of significant events, including natural disasters and environmental crises. At the same time, the government's recognition that environmental risks and harms are of legitimate concern has meant that the state of the environment, like corruption but unlike some other sensitive issues such as Taiwan and the Tiananmen Square crackdown, is now a political issue that is relatively open to public debate. Indeed, some major environmental crises have served to raise public awareness of the scale of the environmental risks faced by China and put pressure on the government to tackle them. When a major earthquake struck Sichuan Province in 2008, citizens used social media to report on and mobilise in response to the disaster. Along with reporting by mainstream media outlets, which largely ignored instructions to suppress or downplay the news once it became apparent that information about the earthquake was spreading rapidly online, this turned the earthquake into a huge media event. In mid-2007, a large algal bloom in Lake Taihu in Jiangsu Province meant that 2.3 million residents in the city of Wuxi were effectively without tap water for a week (Chen 2009: xxi), creating a social media storm. Around the same time, residents of Xiamen in Fujian Province were using their mobile phones to send hundreds of thousands of text messages to organise protests against the planned opening of a paraxylene plant that activists claimed would pose a serious risk of harm to the health of local residents (Cody 2007). The 'Xiamen PX' protests, as they became known, eventually resulted in the relocation of the plant and were followed in 2011 by another protest with the same outcome in Dalian after reports of storm damage to a dyke designed to protect a local paraxylene facility brought the existence of the plant to the attention of concerned residents (Tang 2011).

In response to the growing difficulty of preventing knowledge of an event from reaching the public, the authorities have increasingly turned to more sophisticated and proactive strategies designed to shape public understanding of an issue or event by controlling the way it is framed in the news media, rather than simply suppressing information. The long-term shift has been from a fairly crude form of ideological indoctrination to a more subtle focus on dominating public discourse (Zhao 1998: 10).

Japan

The history of Japan has been shaped by its status as a late-comer industrial power seeking to catch up with the West. The opening of the country following the arrival of US Commodore Matthew Perry's squadron of 'black ships' in 1853 set in motion a domestic struggle between samurai aiming to restore the emperor and modernise the country and opponents resisting the encroachment of foreign powers. The victory of the modernisers meant a change in leadership from the ruling shogunate to the emperor and the establishment of the Meiji government in 1868. In the face of the Western imperial expansion into East Asia, the new government adopted a policy of *fukoku kyōei* ('enrich the nation and strengthen the military'). The *kyōei* aspect of the policy was integral to Japan itself soon becoming an imperial power. This engendered a range of risks for the population, most obviously the Sino-Japanese war of 1894–1895 and the Russo-Japanese war of 1904–1905. Meanwhile, the *fukoku* part of the policy led to rapid industrialisation and a commitment to economic growth, but the lack of regulation gave rise to a range of risks to the population from pollution to wider environmental degradation. In this way, human-induced harm to the environment and accompany risks and harm for the population were intertwined with the national policies Japan pursued as a late-comer industrial power.

Prioritising rapid industrial development over the environment entails costs. Illustrative of the risk of harm to the population in the late nineteenth century is the environmental destruction caused by copper mining. Copper played a particularly important role for the Meiji government as its export generated essential foreign currency for the purchase of military equipment as well as the industrial machinery indispensable to development and modernisation. But copper mining was a double-edged sword: a source of foreign currency, on the one hand, but also a source of pollution, on the other. Industrial waste from the mining polluted waterways and was the source of the harm to the health of local residents. This meant that environmental concerns in Japan were focused from the start on the harm to the population caused by industrial pollution (McKean 1981: 136).

The first major case of pollution to capture media headlines was the Ashio Copper Mine, which produced the most copper ore in Japan in the late nineteenth century. The environmental damage from the mine's copper-ore slurry included pollution of waterways, the killing of fish and increased mortality rates among the population; flooding and contamination of 1,600 hectares of farmland; and the flooding of villages in Tochigi and Gunma prefectures with contaminated water (Shoji and Sugai 1992; Stolz 2014). The risk posed by the pollution led to protests by farmers and others to put pressure on the prefectures and, as time passed, the national government to take responsibility in regulating the mine. The initial aim of the protestors was to close down the mine in order to eliminate the risk posed, but gradually their aim changed into making the company responsible for offering compensation for the harm caused (Shoji and Sugai 1992).

The news media played a crucial role in spreading information about such incidents of pollution, but as became clear as the protests against the mine developed over the years, the media did not always direct public opinion against the pollution but at times supported the state and big business against the protestors. In the case of the Kawamata incident of 1900, for instance, several thousand farmers making their way to Tokyo to protest against the pollution caused by the Ashio mine were subject to strong-arm tactics by the police, with seventy arrested and charged with rioting in Kawamata under the new Formenting Rebellion Act (Tsuru 2012: 35). The anti-pollution activist and Diet member Tanaka Shōzō was crucial in promoting these political protests against the mine and taking the problem of responsibility for the pollution to the highest authority in the land when in 1901 he appealed to the emperor to address the issue (Stolz 2014). However, this did not lead to the media standing on the side of the farmers (Shoji and Sakai 1992).

This history of popular protests going back to the case of the Ashio mine is the background for the numerous post-war citizens' movements launched against a range of incidents of pollution posing the risk of harm and actual harm to the local population. The priority the government continued to place on rapid industrialisation and economic growth following the defeat and devastation of the war meant that, until the early 1970s, policymakers tended to side with the producers rather the victims of pollution. At the same time, though, citizens not only mounted political protests against the source of the pollution, as in the case of those against the Ashio mine, but also often sought recourse in the courts to establish responsibility for the risks and harm caused by pollution and environmental degradation.

The most important cases in the post-war era are the so-called big four pollution cases, which were brought before the courts between 1967 and 1969 and the verdicts issued between 1971 and 1973. The 'big four' occurred at a time of a major increase in pollution as well as a growing role for the media in disseminating information about its harmful effects on health. As an illustration, 'in 1971 sulfur dioxide emissions in the three major municipal areas of Japan [Tokyo, Osaka, Nagoya] were three times the national average. . . . On some days it was not possible to go outside at all' (Japan International Cooperation Agency n.d.). With the support of the media, environmental protestors continued to put pressure on the government to take responsibility for implementing effective legislation to combat pollution in what has been called 'Japan's long environmental sixties' (Avenell 2012a: 425). This bore fruit at the subnational level, as in the case of the implementation of pollution controls and support for the environment by progressive governors such as Minobe Ryōkichi in Tokyo (1967–1979 incumbent) (Tokyo Metropolitan Government 1977; McKean 1981). It also bore fruit at the national level, when in July 1970 grassroots efforts eventually led the government to establish the Headquarters for Countermeasures for Environmental Pollution and pass fourteen environmental bills in an extraordinary session of the Diet (the so-called Pollution Diet) establishing responsibility for the pollution under the polluter pays principle (Larson 2005: 561–563). The Diet session was followed by the creation in 1971 of the Environment Agency (Schreurs 2002: 45–46). This raft of anti-pollution legislation was at a pace unseen in the world before (Broadbent 1998).

The 'big four' cases demonstrate how the Japanese government's prioritisation of rapid industrialisation over the environment led to a range of harmful effects on the health of the population as a result of pollution. The first major case to gain media attention is that of poisoning in Toyama Prefecture caused by the discharge of cadmium into a local river by the Mitsui Mining and Smelting Company. This discharge, which had occurred from 1910 onwards, poisoned the river basin and drinking water as well as local food, giving rise to *itai-itai* ('ouch-ouch') disease (Kasuya 2000). The severe pain in the joints and spine suffered by the victims gave rise to the disease's name. It was not until 1968, however, that a number of the victims brought a lawsuit against Mitsui and, after the end of the appeal process, the company took responsibility for the pollution by paying compensation to a small number of plaintiffs (i.e. not all of the victims), albeit on court orders.

The second case is Minamata disease, named after the location where fish poisoned by methylmercury were consumed by the local population in the fishing village of Minamata, Kumamoto prefecture. The contamination of Minamata Bay was traced to the discharge of industrial waste containing mercury by the Chisso Corporation. The disease, first identified in 1956, leads to a deterioration in the victim's central nervous system. It results in a range of symptoms such as difficulty in walking, loss of sensation and convulsions and eventual death (Tsuda *et al.* 2009). Protests by the victims and their supporters were combined with legal battles and success in the courts. The company was eventually forced to take responsibility for financial compensation to thousands of the victims following these endeavours. It was the largest settlement in Japan at the time the case was won. However, the efforts made to tackle the pollution were often frustrated by the resistance of those whose livelihood depended on Chisso, who were reluctant to challenge their employer in what was in essence a 'company town'.

A later incident of Minamata disease occurred in 1965 in Niigata prefecture. This third of the 'big four' pollution cases broke out as a result of industrial waste containing mercury being discharged into a local river by Showa Denko. The Showa Denko factory was similar to Chisso in discharging mercury from the industrial process. This meant that experienced experts from the earlier case were able to identify the company as the cause of the contamination. Unlike in the case Kumamoto, where the local population was divided due to the role of Chisso as the company in the 'company town', locals backed the victims in protesting against the company. A successful lawsuit was launched in 1968. Indeed, the struggle of the victims in Niigata led to the victims in Kumamoto pursuing their own lawsuit for compensation (George 2002). Later, these cases of mercury poisoning in Japan became the inspiration for the 2013 global convention on the reduction of mercury in the environment held in Minamata, Kumamoto. It was adopted by 139 governments and signed by 92, including Japan (United Nations Environmental Programme 2013).

The final case is the pollution caused by the sulphur dioxide and other harmful contaminants pumped into the atmosphere at the Yokkaichi petrochemical complex in Yokkaichi, Mie prefecture. The complex produced over a quarter of Japan's total petroleum production during the 1960s, when widespread cases of asthma (Yokkaichi asthma) came to the media's attention (Imai *et al.* 1981). In the wake of popular protests and subsequent court cases brought by the victims against Shōwa Yokkaichi Sekiyū and other companies, the media covered the cases at both the national and local levels. It was based on the experience of Yokkaichi that, in order to protect public health, legislation was passed aimed at regulating the amount of harmful contaminants discharged into the atmosphere.

A widespread media campaign served to build the support needed to tackle environmental degradation during the era of the 'big four' pollution cases. Although the initial response of the media was to ignore the anti-pollution movement (Avenell 2012b: 245), they later came to play a crucial role in promoting 'public awareness of the deteriorating environmental conditions across the country' (Imura 2005: 26–27). This can be seen, for instance, in the print media in the case of the *Asahi Shimbun*, which created a media team to report on pollution issues (Avenell 2012a: 431). The media also played a crucial role in raising awareness by covering the testimonies made by the victims in the big four pollution trials, which 'awakened the public to the dangers of pollution and thus contributed to the creation of citizens' movements' (McKean 1981: 79). In this way, following protest by the victims and widespread coverage of the pollution cases by the media, along with pressure from progressive subnational political authorities, the central government gradually began to change tack from economic growth and modernisation at all costs to establishing responsibility under the polluter pays principle, implement countermeasures to regulate industry and protect the environment. These efforts bore fruit in a range of areas, especially in cleaning up air pollution, which citizens' movements and subnational political authorities such as Tokyo had long campaigned to make a reality.

Japan's rapid cleanup of pollution and the environmental regulations put in place from the 1970s onwards has captured global headlines over the years as the image of Japan as a 'toxic archipelago' has faded (Walker 2010). However, the explosion and partial meltdown at the Fukushima Daiichi nuclear plant at the time of the March 2011 (3.11) Great East Japan earthquake and tsunami has brought to the fore once again the balance between economic growth and responsibility for protection of the environment. Due to the crucial role of the media in disseminating information on environmental issues, reportage on the 'big four' pollution cases alerted the public to the way the government at first supported industry rather than the victims of pollution, although in time the government bowed to pressure from the grassroots. However, the government's ongoing response to the Fukushima disaster, the continuation of radiation leakage and revelations about information not being provided to the public earlier, brings into focus the role the media has played in the nuclear crisis.

The government's adoption of nuclear power as a source of energy has often faced anti-nuclear sentiment in the media and on the grassroots level. While the atomic bombing of Hiroshima and Nagasaki at the end of World War II helps to explain the resistance to the use of nuclear power, there were strong objections to the location of nuclear power stations in communities due to concern over the risk to the environment as well, prompting the government to seek more amenable communities for the location of nuclear reactors (Aldrich 2010). At the same time, though, during the late 1960s and early 1970s opinion surveys demonstrated a large number of people supported the continued building of nuclear power plants (Aldrich 2013: 255). This changed in the wake of nuclear accidents like Three Mile Island and Chernobyl. While protests against nuclear power had occurred for many years before 3.11, seldom were they on the scale of those taking place in the wake of the Fukushima Daiichi meltdown. Tens of thousands took to the streets to protest against nuclear energy and call for the closing down of nuclear reactors (Setouchi *et al.* 2012). Indeed, the Fukushima disaster has brought about a major change in the public's attitude to nuclear energy, with a majority now wanting change to the role of nuclear power in Japan's energy mix (Aldrich 2013: 255–258). This can be seen, for instance, in the results of public opinion surveys before and after 3.11 (Table 2.1).

Table 2.1 Japanese polls: 'What should be done about the future of nuclear power?'

	Increase	*Maintain status quo*	*Decrease**	*Abolish**	*Unknown*
December 2005	55.1	20.2	14.7	2.3	7.7
October 2009	59.6	18.8	14.6	1.6	5.4
4 April 2011	10	46	29	12	3
18 April 2011	5	51	30	11	3
18 April 2011	4.2	48.5	33.3	10.5	3.5
16 May 2011	4	34	44	15	3
9 June 2011	4	41	36	16	3
3 July 2011	2	29	46	19	4
5 August 2011	5	49	32	13	1
28 October 2011	2	23	42	24	9

* The Prime Minister's Office surveys have been recoded into the categories of 'decrease' and 'abolish' to match the wording of the later newspaper surveys.

Sources: December 2005 and October 2009 public opinion data from Prime Minister's Cabinet Office surveys; 4 April 2011 from *Yomiuri Shimbun* (Yomiuri Newspaper); 18 April and 28 October 2011 data from NHK poll; 18 April, 16 May, 9 June 2011 from Asahi Newspaper; 3 July 2011 data from Yomiuri Newspaper; and 5 August 2011 data from Soka Gakkai poll.

From Aldrich (2013: 256). Also see table in Hasegawa (2014: 296).

In the face of public protest, most of Japan's reactors were shut down for safety inspections by early in 2012 and all of them by September 2013. The responsibility to restart nuclear power generation was put in the hands of local governments, but the local populations at the site of the reactors have remained generally opposed to the restart, anxious over the risks posed. As time has passed, however, the present Abe Shinzō government has pushed for the restart of the downed reactors, and certain localities have been more supportive of a restart than others (Kingston 2013). For instance, Kyushu Electric Power Company's Sendai 1 reactor was back in full operation in late 2015 after passing the increased safety standards put in place after the Fukushima disaster and following approval by the local authorities.

The impact of the Fukushima meltdown on health is the most salient example of the risk posed to the local population by nuclear reactors. While research is still ongoing, the worldwide human impact has been estimated at approximately 130 deaths and 180 cancers, overwhelmingly in Japan. This excludes the risk of harm to the health among the approximately 20,000 workers at the plant after the accident (Ten Hoeve and Jacobson 2012). While such figures will remain controversial, the disaster also has exerted a profound effect on the way mothers remain still concerned about the risk to the health of their offspring in Mie prefecture and further afield, as '[t]hey no longer trust official government and media reports on "safety"' (Hasegawa 2014: 293).

At the same time, the question of trust in the media is related to the media landscape in Japan. The ownership structure means that the commercial radio and television stations are linked to the major print media. Thus, Japan's major newspapers are majority stakeholders in their affiliate networks of Nihon TV, TBS, Fuji TV and TV Asahi. With the ideological spectrum of these newspapers running from the *Sankei Shimbun* on the right and the *Asahi Shimbun* on the left, trust in news sources is linked to the ideological orientation of the news sources. Even the public broadcaster, NHK (*Nihon Hōsō Kyōkai*, Japan Broadcasting Corporation), is now under constraint following the decision of Prime Minister Abe to hand-pick the new director general of NHK in December 2013, who takes the position that the corporation's international news programmes should toe the government line (Soble 2014). In the case of nuclear power, moreover, a veteran commentator on NHK radio quit in protest at not being able to discuss nuclear power on air until after the upcoming February 2014 gubernatorial election in Tokyo (Otake 2014).

Another way trust in the media is constrained is through the existence and role of the press clubs, which shape the way journalists interact with the sources for their stories and help to set the media agenda (Freeman 2000). In the case of the Fukushima Daiichi meltdown, for instance, as freelance journalists are not members of the press clubs they are excluded from press club meetings with government ministers. The point is illustrated by the case of the members of the press who were permitted to accompany Hosono Gōshi, the minister in charge of nuclear power, on an inspection tour inside the reactor area shortly after the accident. Freelance journalists were excluded (Hatakeyama 2015).

In short, the press clubs effectively act as 'information cartels' in determining access to stories (Freeman 2000).

The close relationship and vested interests these information cartels instil between the media, policymakers and the nuclear industry gave rise to the so-called nuclear village; that is, the pronuclear government, companies and academia (Sugiman 2014). This has meant that, rather than the media going to the heart of the problem about the risk posed to safety by nuclear power, 'any problems with a nuclear power plant were good materials to criticise the electric power companies. But, the criticism was concerned only with attracting readers and audiences. It was rare for the mass media to discuss seriously the details of possible severe accidents and their subsequent social disaster' (Sugiman 2014: 263). However, in the wake of the nuclear disaster at Fukushima Daiichi, there has been some soul-searching in the media, and reporters are less sanguine about safety and more aware of the risks of harm posed by nuclear power (Asahi Shimbun 2013). Given the diverse ideology of the news sources, no consensus exists in the media on the Abe administration's policy of restarting the nuclear reactors, with newspapers such as the *Sankei* in favour and the *Asahi* more sceptical of the government's pro-restart nuclear policy.

Apart from ideological orientation, media coverage of the Fukushima disaster also brought into focus the degree of objectivity and emotionality in media coverage of the accident. In a study of the media following the disaster from March until September 2011, the 'news media generally reported neutral and objective factual information' on the accident, although journalists who were surveyed afterwards were affected in their evaluation by their own negative experience at the time of the accident (Uchida *et al.* 2015). Other work has demonstrated how media reports on the accident are entwined with health-related anxieties. In this case, the source of the media has an impact on the degree of anxiety of the consumer, with radio being seen to provide an appropriate medium to help reduce anxiety in a post-disaster environment (Sugimoto *et al.* 2013). Finally, a case study of the prefectural newspaper in Aomori, the *Tōōnippō*, during the first few months after the earthquake demonstrates how the nuclear disaster at Fukushima was reported at the local level. It shows that, in the early stages, reporting focused on the nuclear power industry rather than the role of the government in dealing with the accident (Rausch 2012).

Conclusion

The above discussion of Australia, China and Japan has sought to set the scene for the case studies in the following chapters, both in terms of the historical context for the emergence of environmental issues and in terms of how risks, responsibilities and pollution are interlinked. We have seen how the three countries at the heart of this study have faced both similar and different environmental risks, with the responsibility for tackling the risks and the manifestation of risks as harm diverging over both spatially and temporally. What the above overview makes clear is that all three countries have faced a range of environmental risks and shoulder the responsibility for addressing these risks in a variety of ways, mediated by government, environmental protest movements and the media.

To start with, we found in the case of China a strong role for the government in taking the responsibility to tackle environmental risks, such as pollution, but not to the total exclusion of other actors and the role of the media in shaping the environmental discourse. This is seen in the way popular opposition emerged to the construction of the Three Gorges Dam, demonstrating how the people can to a certain extent exercise a voice in tackling environmental issues in China. We also saw how GONGOs play a role in tackling environmental risks, thereby bridging the gap between the societal and governmental levels.

In the case of Australia, the role of the environmental movement is similar in some respects to China, as the protests against the risks posed by the damming of the Franklin River demonstrate, but the role of the popular voice and the media in Australia is linked more to the international environmental movement and global trends than it is in China. There have been a large number of environmental protests against the construction of dams in Japan, too, and complaints from local residents have even led to the start of the first removal of a dam, the Arase Dam in Kumamoto (Japan for Sustainability 2014). But rather than protests about Japan's role and position as one of the world's big dam builders, the media and popular protests mean the country is best known for the success of the environmental movement in the 'big four' cases of industrial pollution. Here the main issue was the manifestation of environmental risk as harm to the health of the population, portrayed vividly in the media. The risks posed and the harm inflicted on the population has been mediated by the courts, as protestors used legal means of redress in the cases of *itai-itai* disease, Minamata disease and the harm to health caused by atmospheric pollutants at Yokkaichi. In this way, the protests against the 'big four' led the courts to clarify responsibility under the polluter pays principle.

In the cases of Australia and Japan, we have seen over time how the governments and populations of both countries come to accept that pollution and other forms of environmental degradation are a barrier to economic growth, and are issues that need to be tackled. A distinctive feature of the case in China is the way the media is constrained in challenging the government's continued acceptance of pollution as one of the costs of economic growth for China as a developing country. This means that, while measures have been taken to tackle the risks posed by climate change, the position of China as a developing country means that gaining agreement on sharing responsibility for addressing climate change and other environmental risks at the regional and global levels remains a challenge. Unlike in the cases of Japan and Australia, where international pressure and media have been integral to the way climate change has been addressed, China continues as a major source of environmental risks.

We do need to acknowledge that the integration of the Chinese economy into the global production and consumption chain means that countries like Australia and Japan are to a certain extent 'outsourcing' some of their own environmental risks to China. On the one hand, Australia is a supplier of raw materials needed by the Chinese economy. This not only poses risks to the environment in Australia, through the extraction of resources for export to factories in China, but the production of consumer goods exported to the Australian market cannot be divorced from the environmental risks posed to the Chinese people from those factories. On the other hand, Japanese multinational corporations are using China as a production platform for consumer goods, meaning that the environmental risks posed by industrial waste and atmospheric pollution, once faced by the Japanese population, is now faced by the Chinese. Even more so than Australia, then, Japan is implicated in the production of environmental risks and harms in China. How responsibility for these forms of transnational interlinkages can be allocated remains fraught with difficulty, given the continuing role of the territorial sovereign state as one of the main sources for mediating and governing environmental risk and harm. In this sense, states have not only national and regional responsibilities to tackle environmental risks, but global ones, too.

References

Aldrich, Daniel P. (2010) *Site Fights: Divisive Facilities and Civil Society in Japan and the West.* Ithaca, NY: Cornell University Press.

Aldrich, Daniel P. (2013) 'Rethinking civil society-state relations in Japan after the Fukushima accident'. *Polity* 45(2): 249–264.

Asahi Shimbun (2013) *Genpatsu to Media. 3.11 Sekinin no ari ka.* Tokyo: Asahi Shimbun Shuppansha.

Australian Communications and Media Authority (n.d.) *Media Interests 'Snapshot'.* Accessed on 21 January 2016 from www.acma.gov.au/theACMA/media-interests-snapshot

Australian Government (2012) *Australia in the Asian Century.* Accessed on 16 November 2012 from http://asiancentury.dpmc.gov.au/white-paper

Avenell, Simon (2012a) 'Japan's long environmental sixties and the birth of a green Leviathan'. *Japanese Studies* 32(3): 423–444.

Avenell, Simon (2012b) 'From fearsome activism and the nuclear blind spot in contemporary Japan'. *Environmental History* 17: 244–276.

Baum, Richard (1994) *Burying Mao: Chinese Politics in the Age of Deng Xiaoping.* Princeton, NJ: Princeton University Press.

Bonyhady, Tim (2003) *The Colonial Earth* (Vol. 34). Melbourne: Melbourne University Publishing.

Boyd, Olivia (16 September 2014) 'Human toll of air pollution could be costing China 13% of GDP'. *Chinadialogue.* Accessed on 14 December 2015 from www.china dialogue.net/blog/7316-Human-toll-of-air-pollution-could-be-costing-China-13-of-GDP/en

Brady, Anne-Marie (2008) *Marketing Dictatorship: Propaganda and Thought Work in Contemporary China.* Lanham, MD: Rowman and Littlefield.

Broadbent, Jeffrey (1998) *Environmental Politics in Japan: Networks of Power and Protest.* Cambridge, UK: Cambridge University Press.

Chen, Gang (2009) *Politics of China's Environmental Protection: Problems and Progress.* Singapore: World Scientific.

China Daily (24 October 2007) 'Ecological civilization'. Accessed on 14 December 2015 from www.chinadaily.com.cn/opinion/2007–10/24/content_6201964.htm

Cody, Edward (28 June 2007) 'Text messages giving voice to Chinese'. *Washington Post*. Accessed on 14 December 2015 from www.washingtonpost.com/wp-dyn/content/article/2007/06/27/AR2007062702962.html

Crossley, Pamela Kyle (2010) *The Wobbling Pivot, China since 1800: An Interpretive History*. Chichester, West Sussex: Wiley-Blackwell.

Dai, Qing (1994) *Yangtze! Yangtze!* Patricia Adams and John Thibodeau (eds), translated by Nancy Liu, Wu Mei, Sun Yougeng and Zhang Xiaogang. London: Probe International Earthscan.

Economy, Elizabeth (2007) 'The great leap backward?'. *Foreign Affairs* 86(5): 38–59.

Economy, Elizabeth (2010) *The River Runs Black: The Environmental Challenge to China's Future* (2nd edition). Ithaca, NY: Cornell University Press.

Edmonds, Richard Louis (2011) 'The evolution of environmental policy in the People's Republic of China'. *Journal of Current Chinese Affairs* 40(3): 13–35.

Edney, Kingsley and Jonathan Symons (2014) 'China and the blunt temptations of geo-engineering: The role of solar radiation management in China's strategic response to climate change'. *The Pacific Review* 27(3): 307–332.

Elvin, Mark (2008) *The Retreat of the Elephants: An Environmental History of China*. New Haven, CT: Yale University Press.

Flew, Terry and Ben Goldsmith (8 August 2013) 'FactCheck: Does Murdoch own 70% of newspapers in Australia?'. *The Conversation*. Accessed on 21 January 2016 from http://theconversation.com/factcheck-does-murdoch-own-70-of-newspapers-in-australia-16812

Freeman, Laurie A. (2000) *Closing the Shop: Information Cartels and Japan's Mass Media*. Princeton, NJ: Princeton University Press.

George, Timothy S. (2002) *Minamata: Pollution and Struggle for Democracy in Postwar Japan*. Boston, MA: Harvard University Press.

Gilley, Bruce (2012) 'Authoritarian environmentalism and China's response to climate change'. *Environmental Politics* 21(2): 287–307.

Guardian (12 June 2014) 'A sickening feeling of betrayal: the new battle for Maules Creek'. Accessed on 21 January 2016 from www.theguardian.com/world/2014/jun/12/the-new-battle-for-maules-creek

Hanson, Fergus and Andrew Shearer (2009) 'China and the world: Public opinion and foreign policy'. Sydney: Lowy Institute for International Policy. Accessed on 14 December 2015 from www.lowyinstitute.org/files/pubfiles/Lowy_China_Poll_2009_Web.pdf

Hasegawa, Koichi (2014) 'The Fukushima nuclear accident and Japan's civil society: Context, reactions, and policy impacts'. *International Sociology* 29(4): 283–301.

Hatakeyama, Michiyoshi (2015) *Fukushima daiichi genpatsu. Kisha kurabu gentei kōkai e no teikō*. Accessed on 20 July 2015 from http://hatakezo.jugem.jp/?eid=35

Hay, Peter (1991) 'Destabilising Tasmanian politics: The key role of the Greens'. *Bulletin of the Centre for Tasmanian Historical Studies* 3(2): 60–70.

Held, David, Eva-Maria Nag and Charles Roger (2011) 'The governance of climate change in China'. *LSE Global Governance Working Paper* 1: 22–24.

Herman, Edward S. and Noam Chomsky (2010) *Manufacturing Consent: The Political Economy of the Mass Media*. New York: Random House.

Ho, Peter (2001) 'Greening without conflict? Environmentalism, NGOs and civil society in China'. *Development and Change* 32(5): 893–921.

Ho, Peter (2007) 'Embedded activism and political change in a Semiauthoritarian context'. *China Information* 21(2): 187–209.

Ho, Peter and Richard Louis Edmonds (2007) 'Perspectives of time and change: Rethinking embedded environmental activism in China'. *China Information* 21(2): 331–344.

Imai, Masayuki, Katsumi Yoshida, Yasuko Tomita, Kazuo Kasama, Masayoshi Kitabatake, and Hidehiko Oshima (1981) 'Air pollution levels and death from chronic obstructive lung disease in Yokkaichi Japan'. *Japanese Journal of Hygiene* 36(4): 671–677.

Imura, Hidefumi (2005) 'Japan's environmental policy: Past and future', in Hidefumi Imura and Miranda A. Schreurs (eds) *Environmental Policy in Japan*. Cheltenham, UK: Edward Elgar, pp. 15–48.

Jackson, Sukhan and Adrian C. Sleigh (2001) 'The political economy and socio-economic impact of China's Three Gorges Dam'. *Asian Studies Review* 25(1): 52–72.

Japan International Cooperation Agency (n.d.) *Environmental Pollution Control Measures (chapter 6)*. Accessed on 22 January 2016 from http://jica-ri.jica.go.jp/IFIC_and_JBICI-Studies/english/publications/reports/study/topical/health/pdf/health_08.pdf

Japan for Sustainability (28 November 2014) *Newsletter No.147*. Accessed on 11 December 2015 from www.japanfs.org/en/news/archives/news_id035105.html

Kasuya, M. (2000) 'Recent epidemiological studies on itai-itai disease as a chronic cadmium poisoning in Japan'. *Water Science and Technology* 42(7–8): 147–154.

Kingston, Jeffrey (2013) 'Nuclear power politics in Japan, 2011–2013'. *Asian Perspective* 37(4): 501–521.

Larson, Eric T. (2005) 'Why environmental liability regimes in the United States, the European Community, and Japan have grown synonymous with the polluter pays principle'. *Vanderbilt Journal of Transnational Law* 38(1): 541–575.

Lester, Libby (2005) 'Wilderness and the loaded language of news'. *Media International Australia Incorporating Culture and Policy* 115: 123–134.

Lester, Libby (2007) *Giving Ground: Media and Environmental Conflict in Tasmania*. Hobart: Quintus.

Lester, Libby (2013) 'On flak, balance and activism: The ups and downs of environmental journalism', in Stephen Tanner and Nick Richardson (eds) *Journalism Research and Investigation in a Digital World*. Melbourne: Oxford University Press, pp. 221–231.

Lester, Libby (2014) 'Transnational publics and environmental conflict in the Asian century'. *Media International Australia* 150: 167–178.

Lester, Libby and Brett Hutchins (2009) 'Power games: Environmental protest, news media and the internet'. *Media, Culture & Society* 31(4): 579–595.

Lester, Libby, Lyn McGaurr, and Bruce Tranter (2015) 'The election that forgot the environment? Issues, EMOs, and the press in Australia'. *The International Journal of Press/Politics* 20: 3–25.

Li, Jing (28 March 2013) '1.1 trillion yuan in economic losses from pollution in 2010, China Report says'. *South China Morning Post*. Accessed on 14 December 2015 from www.scmp.com/news/china/article/1201364/11-tr-yuan-economic-losses-pollution-2010-china-report-says

Li, Jing (7 May 2015) 'China to get tougher on eco-unfriendly officials'. *South China Morning Post*. Accessed on 14 December 2015 from www.scmp.com/news/china/policies-politics/article/1787456/china-impose-tougher-punishments-officials-found-cause

Li, Vic and Graeme Lang (2010) 'China's "green GDP" experiment and the struggle for ecological modernisation'. *Journal of Contemporary Asia* 40(1): 44–62.

Lieberthal, Kenneth (1992) 'The "fragmented authoritarianism" model and its limitations', in Kenneth Lieberthal and David M. Lampton (eds) *Bureaucracy, Politics, and Decision Making in Post-Mao China*. Berkeley, CA: University of California Press, pp. 1–30.

Lohrey, Amanda (2002) 'Groundswell: The rise of the greens'. *Quarterly Essay* 8: 1–86.

Lora-Wainwright, Anna (2010) 'An anthropology of "cancer villages": Villagers' perspectives and the politics of responsibility'. *Journal of Contemporary China* 19(63): 79–99.

McGaurr, Lyn, Libby Lester, and James Painter (2013) 'Risk, uncertainty and opportunity in climate change coverage: Australia compared'. *Australian Journalism Review* 35(2): 21–34.

McGrath, Matt (21 September 2014) 'China's per capita carbon emissions overtake EU's'. *BBC News*. Accessed on 14 December 2015 from www.bbc.com/news/science-environment-29239194

McKean, Margaret A. (1981) *Environmental Protest and Citizen Politics in Japan*. Berkley, CA: University of California Press.

McKnight, David (2010) 'A change in the climate? The journalism of opinion at News Corporation'. *Journalism* 116: 693–706.

Meadows, Michael and Robert Thomson (2013) 'Campaigning journalism: The early press, environmental advocacy and national parks', in Libby Lester and Brett Hutchins (eds) *Environmental Conflict and the Media*. New York: Peter Lang, pp. 37–48.

Minter, Adam (2013) *Junkyard Planet: Travels in the Billion-Dollar Trash Trade*. New York: Bloomsbury.

Moore, Scott M. (2014) 'Modernisation, authoritarianism, and the environment: The politics of China's south-north water transfer project'. *Environmental Politics* 23(6): 947–964.

Muscolino, Micah S. (2015) *The Ecology of War in China: Henan Province, the Yellow River, and Beyond, 1938–1950*. Cambridge, UK: Cambridge University Press.

Otake, Tomoko (30 January 2014) 'Scholar quits NHK over nuclear power hush-up'. *Japan Times*. Accessed on 14 July 2015 from www.japantimes.co.jp/news/2014/01/30/national/scholar-quits-nhk-over-nuclear-power-hush-up/#.Va-RHUYaOzx

Rausch, Anthony (2012) 'Framing a catastrophe: Portrayal of the 3.11 disaster by a local Japanese newspaper'. *Electronic Journal of Contemporary Japanese Studies* 12(1). Accessed on 15 July 2015 from www.japanesestudies.org.uk/ejcjs/vol12/iss1/rausch.html

Reuters (9 July 2013) 'China official defends "embarrassing" environmental protection ministry'. *Reuters.com*. Accessed on 14 December 2015 from www.reuters.com/article/2013/07/09/us-china-environment-idUSBRE9680A920130709

Ryan, Fergus (26 September 2015) 'China and US agree on ivory ban in bid to end illegal trade globally'. *The Guardian*. Accessed on 14 December 2015 from www.theguardian.com/environment/2015/sep/26/china-and-us-agree-on-ivory-ban-in-bid-to-end-illegal-trade-globally

Schreurs, Miranda A. (2002) *Environmental Politics in Japan, Germany and the United States*. Cambridge, UK: Cambridge University Press.

Setouchi, Jakuchō, Kamata Satoshi, and Karatani Kōjin (2012) *Datsu Genpatsu to Demo: Soshite, Minshushugi*. Tokyo: Chikuma Shobō.

Shapiro, Judith (2001) *Mao's War against Nature: Politics and the Environment in Revolutionary China*. Cambridge, UK: Cambridge University Press.

Shoji, Kichiro and Masuro Sugai (1992) 'The Ashio copper mine pollution case: The origins of environmental destruction', in Jun Ui (ed) *Industrial Pollution in Japan*. Tokyo: United Nations University Press. Accessed on 8 July 2015 from http://archive.unu.edu/unupress/unupbooks/uu35ie/uu35ie04.htm

Soble, Jonathan (4 February 2014) 'Abe's interference blurs the picture at Japan's NHK broadcaster'. *Financial Times*. Accessed on 14 July 2015 from www.ft.com/cms/s/0/33e00cbe-8d72-11e3-9dbb-00144feab7de.html#axzz3gcjN0fuC

Spence, Jonathan (1990) *The Search for Modern China*. New York: W.W. Norton and Company.

Stolz, Robert (2014) *Bad Water: Nature, Pollution, and Politics in Japan, 1870–1950*. London and Durham, NC: Duke University Press.

Sugiman, Toshio (2014) 'Lessons learned from the 2011 debacle of the Fukushima nuclear power plant'. *Public Understanding of Science* 23(3): 254–267.

Sugimoto, Anima, Shuhei Nomura, Masaharu Tsubokura, Tomoko Matsumura, Kaori Muto, Mikiko Sato, and Stuart Gilmour (2013) 'The relationship between media consumption and health-related anxieties after the Fukushima Daiichi nuclear disaster'. *PLoS ONE*, 8: 8. Accessed on 20 July 2015 from http://journals.plos.org/plosone/article?id=10.1371/journal.pone.0065331

Tang, Hao (6 September 2011) 'Public storm in Dalian'. *Chinadialogue*. Accessed on 14 December 2015 from www.chinadialogue.net/article/4511-Public-storm-in-Dalian

Ten Hoeve, John E. and Mark Z. Jacobson (2012) 'Worldwide health effects of the Fukushima Daiichi nuclear accident'. *Energy & Environmental Science* 5: 8743–8757.

Tokyo Metropolitan Government (1977) *Tokyo Fights Pollution* (revised edition). Tokyo: Tokyo Metropolitan Government.

Tranter, Bruce and Libby Lester (2015) 'Climate patriots?: Concerns over climate change and other environmental issues in Australia'. *Public Understanding of Science* (online first). doi:10.1177/0963662515618553

Tsuda, Toshihide, Takashi Yorifuji, Soshi Takao, Masaya Miyai, and Akira Babazono (2009) 'Minamata disease: Catastrophic poisoning due to failed public health response'. *Journal of Public Health Policy* 30(1): 54–67.

Tsuru, Shigeto (2012) *The Political Economy of the Environment: The Case of Japan*. London and New York: Bloomsbury Academic.

Uchida, Yukiko, Chie Kanagawa, Ako Takenishi, Akira Harada, Kiyotake Okawa, and Hiromi Yabuno (2015) 'How did media report on the Great East Japan Earthquake? Objectivity and emotionality seeking in Japanese media coverage'. *PLoS ONE* 10: 5. Accessed on 20 July 2015 from www.ncbi.nlm.nih.gov.eresources.shef.ac.uk/pmc/articles/PMC4436106/?report=classic

United Nations Environmental Programme (2013) *New Global Treaty Cuts Mercury Emissions and Releases, Sets Up Controls on Products, Mines and Industrial Plants*. Accessed on 16 July 2015 from www.mercuryconvention.org/News/Newglobaltreatycutsmercuryemissions/tabid/3470/

Walker, Brett L. (2010) *Toxic Archipelago: A History of Industrial Disease in Japan*. Seattle, WA: University of Washington Press.

World Bank (2015) *The Little Green Data Book*. Washington, DC: World Bank.

Yang, Xi (10 March 2008) 'SEPA gets stronger'. *China.org.cn*. Accessed on 14 December 2015 from www.china.org.cn/environment/news/2008–03/10/content_12143406.htm

Zhan, Jiang (2011) 'Environmental journalism in China', in Susan L. Shirk (ed) *Changing Media, Changing China*. Oxford, UK: Oxford University Press, pp. 115–127.

Zhang, Joy Y. and Michael Barr (2013) *Green Politics in China: Environmental Governance and State-Society Relations*. London: Pluto Press.

Zhao, Yuezhi (1998) *Media, Market, and Democracy in China: Between the Party Line and the Bottom Line*. Urbana, IL: University of Illinois Press.

3 Transnational discourses of risk and responsibility from Australia

Coal, pollution and the Great Barrier Reef

Australia is the first of our case studies. In one way at least, Australia represents the beginning of our transnational pollution story – even if the phenomena we are exploring in this book are best viewed as a circular flow of political communications, goods and environmental harms in which all three countries in this study are caught. Each country is affected by the massive quantities of pollutants emitted daily into the world's atmosphere, and each struggles to control the harmful effects of these pollutants. Each also contributes to the production and circulation of pollution, and to the international politics and communications that inevitably follow. But this is where Australia provides a useful point for us to enter the flow, as not only is it one of the world's largest greenhouse gas emitters per capita, but it also makes a significant contribution to the overall production of greenhouse gases and air pollution by being one of the world's biggest exporters of coal. As a major producer of greenhouse gases, the production and export of coal cannot be separated from climate change, and here Australia provides another point of focus. The Great Barrier Reef, probably Australia's most iconic place, faces serious threats from warming seas and extreme weather events, and has become a global symbol for the impact of continuing pollution and the climate change that results.

This chapter, therefore, analyses political discourses within media in an attempt to reveal Australia's understanding of its role in producing atmospheric pollution, and in taking or attributing responsibility for its impacts. It also considers global reactions and responses to the crisis facing the Great Barrier Reef. We begin by outlining Australia's historic reliance on and future hopes for coal, and then we consider the role of Australia's neighbours and trading partners within major coal developments in Australia, including Japanese and Chinese corporate investment and as major markets. After addressing the role of coal in generating harmful atmospheric pollution, our focus turns to the Great Barrier Reef, which has faced direct impact from the coal industry through proposed dredging to expand coal export facilities and indirect impacts from climate change through warming seas and extreme weather events. We draw on qualitative and quantitative computer-assisted textual analysis to identify and analyse media discourses of risk, responsibility and blame. Who is causing the crisis, who is affected, who is responsible and who responds?

The producer and the affected

The past

Colonised by the British in the late eighteenth century in part as a solution to its overflowing city gaols and the prison hulks lining the Thames, Australia is a product of the Industrial Revolution. Just as it was doing at 'home', coal was to play an important role in the new settlement. Coal is Australia's oldest export trade, and has consistently been one of its biggest. Even Captain James Cook, the British explorer whose journeys in the 1770s provided the impetus and charts required for white settlement in 1788, had learned his trade on the vessels carrying coal between northern England and London (Minerals Council of Australia n.d.), apprenticed to the Walker family, members of which would later move to the new colony themselves to capitalise on the emerging coal industry. The presence of coal was recorded soon after settlement, at the mouth of a river initially named Coal River, but later renamed the Hunter. Australia's first industrial town, Newcastle, is at the centre of the region – its English namesake was at the time Europe's fastest-growing city, a growth created by coal (Hood 2009: 17) – and the Australian Newcastle still retains the biggest coal export facility in the world (Minerals Council of Australia n.d.).

Although early Australian coal mines tended to be short-lived – markets were driven by domestic heating requirements and the whaling industry, and demand was insufficient and too sporadic to sustain activity – the industry was firmly established by the 1840s when a combination of steamships, free settlers and industrial machinery drove growth. No place better illustrates Australia's difficult transition from penal colony to modern industrial nation than the historic coal mines at Saltwater River on the Tasman Peninsula in the southern island state of Tasmania. Hidden in the bush of the remote World Heritage–listed site are rusted remnants of steam-driven mining machinery that (inefficiently) pumped water from the underground pits in which British and Irish convicts who had reoffended post-transportation were forced to work (Australian Government n.d.).

Coal emerged as the driver of Australian growth. It powered sawmills and flour mills, as well as the steamships that brought settlers and goods. With Newcastle and the Hunter Valley housing one of the largest coal fields in the world, the colony (now state) of New South Wales dominated Australia's coal production for the first century and a half of white settlement. In the Hunter region alone, 6.5 million tonnes of coal were mined in 1908, valued then at 2.6 million sterling (Australian Bureau of Statistics 2012). Further north, coal mining began in Queensland shortly after the white settlement of Moreton Bay (near what is now the state capital of Brisbane) and has been continuing ever since. In 1908, Queensland's coal production stood at almost 700,000 tonnes, and estimates of coal deposits across the vast state suggested that 'practically no limit can be set to its possibilities of extension' (Australian Bureau of Statistics 2012).

According to Ian Dunlop, a former fossil fuel executive and now advocate for climate change action:

> This was the core power generation system that the country developed on. . . . If we hadn't had coal, Australia would not be anywhere near what it is today. So it's been absolutely fundamental to the development of the country, particularly in the early years. Had we not had coal, both from a domestic power generation and steel manufacturing perspective and also from an export perspective increasingly since World War II, we would have nowhere near the amount of wealth generation Australia has enjoyed over recent decades.
>
> (in Quince 2015)

Export of Australian coal began within decades of the establishment of the first mines. An American schooner carried 250 tonnes of coal to Rio de Janeiro as early as 1824. The initial poor quality left Asian markets unenthusiastic. With the introduction of steamships, however, Australian coal was transported to a number of refuelling locations in the Asia-Pacific region, including Singapore, Colombo and Suva, and by 1861 the *Sydney Morning Herald* reported: 'We have already despatched our coals to China, Batavia, India, California, and South America. . . . But this is a mere fragment of the probable demand, as the mining interest becomes stable and reliable' (Diamond 2009: 31). In 1865, approximately one-quarter, or 64,000 tonnes, of coal produced in Newcastle was exported.

The export trade was well established by 1910, strongly supported by government. As an *Australian Year Book* of the period notes: 'At present coal mining in Queensland is in a very satisfactory position, the increasing volume of trade being chiefly due to the action of the Government in granting concessions to vessels coaling at local ports' (Australian Bureau of Statistics 2012). Industrial disputes over pay, safety, working conditions and job security impacted the Australian industry heavily in the following decades, but government intervention via the establishment of the Joint Coal Board and then a conciliation labour tribunal allowed rapid expansion post World War II. In the decades following the war, the industry was able to deliver enough coal to support Japan's rise as the global leader in steel production (Quince 2015). Queensland and New South Wales coal drove the expansion, and Australia's economic reliance on trade with its Asian neighbours and their growth became firmly established. Japan and Australia signed their first trade agreement in 1957 (Diamond 2009: 42).

According to Dunlop, Australia was exporting 1.9 million tonnes of coal by 1960, with 96 per cent of these exports for Japan's steel industry (Quince 2015). By 1986, with Queensland production now overtaking New South Wales (Asafu-Adjaye 2009: 46), Australia was exporting 95.7 million tonnes of black coal, of which half was used in steel production and half in generating power (Quince 2015). Mining methods also shifted through this period from underground pits to lower cost and more efficient open-cut mines, in which the 'overburden' was first removed, providing easier access to the coal deposit. Says Dunlop:

> It started to become apparent that the economics of doing that in the '50s and '60s were actually considerably better at scale than underground mining. So whilst you saw underground mining expand in places like the Illawarra in supplying Japan, you then also saw new techniques brought into the Queensland coalfields, particularly in the Bowen Basin, with these open-cut techniques, with very big drain lines, very large-scale trucks – all the pictures you see these days.
>
> (in Quince 2015)

Developing massive deposits for export was expensive, beyond the capacity of the Australia industry alone, and as such companies sought international investments. In 1959, for example, the Australian family company Thiess Bros discovered reserves in Queensland of hard metallurgical (coking coal used in steel production, as opposed to thermal coal used for energy generation) and partnered with Japanese trading company Mitsui and American firm Peabody to develop the deposit, forming Thiess Peabody Mitsui. This company would dominate coal production for the next decade (Hood 2009: 6). While Japanese investment in Australian coal mining continued through the 1960s, it was formalised in the 1970 Ministry of International Trade and Industry (MITI) 'development-for-import' policy (Colley 1997: 1017). Resource security through direct investment in large-scale foreign ventures dispersed across many regions was at the heart of the policy, reflecting Japanese business and government anxiety about the country's vulnerability to resource shortages. As a MITI white paper of 1981–1982, noted:

> Mineral resources are the lifeblood which sustains the life of the people and their industrial activity. Japan depends almost entirely on imports for its mineral resource requirements. What is more, deposits of mineral resources are concentrated in a few areas of the world. . . . Owing to this peculiar set of circumstances, the availability of mineral resources could pose a short-term sporadic threat or a protracted industrial menace to the economic security of Japan.
>
> (in Colley 1997: 1017)

While oil was initially the focus of the policy, coal's wide distribution across a diverse range of regions – regions that, unlike with oil, were without major conflicts – made it an attractive target. Despite Australia's ongoing anxiety about foreign ownership, by 1992–1993 40 per cent of Australian coal production was owned by foreign-based companies, with Japan's stake of 13.1 per cent making it the source of more investment than any other single nation (Colley 1997: 1014). Security of supply – not profits – was the principal stated motivation. Japanese companies did not seek vertical integration within the Australian industry – that is, control and ownership across the full supply chain – and, as such, were able to minimise political risk while providing the concessional financing required for the industry in Australia to rapidly expand through the latter part of the twentieth century. Control remained with local companies, although the Japanese trading houses – despite their minority share of ownership – were able to exert considerable influence through their provision of cheap finances and access to markets (Colley 1997: 1018).

With Japan as its major trading partner and substantial investor, Australia dominated international coal trade until the 1990s, when Indonesia – and for a short period, China – developed thermal coal export industries (Quince 2015). In 2015, Indonesia was the world's largest coal exporter, although its deposits were forecast to run out within thirty years. If predictions are correct, Australia will again become 'coal king' as it rapidly overtakes Indonesia, whose lower-quality coal has already fallen out of favour with China. According to Australia's peak mining industry body, Australia will again be the world's biggest exporter of coal by 2017 (Minerals Council of Australia 2015: 7). China has joined Japan as a major investor in the Australian industry, with overall foreign investment now estimated at 80 per cent of the industry (Quince 2015).

The future

Despite slowing markets, the coal industry remains one of Australia's most lucrative – second only to iron ore in terms of export earnings. Australia holds 9.2 per cent of the world's coal deposits, and almost 80 per cent of the coal dug from these deposits is shipped to its major buyers: Japan, which bought 120 million tonnes of the total 527 million tonnes extracted in 2014, then China, South Korea and India. According to the Minerals Council of Australia (2015), in the five years from 2010, coal accounted on average for more than 15 per cent of Australia's total exports, with export earnings at A\$38.6 billion equal to or exceeding Australia's total agricultural exports. It estimates that the coal industry employs more than 40,000 Australians and another 135,000 indirectly.

The Australian government and mining industry publicly present a rosy picture for the future of coal exports. They predict an increase of 65 million tonnes in the decade to 2025, citing fifty-three proposed new developments for coal mines. They suggest that export value will grow at a rate of 4.8 per cent annually until 2020. While 2015 was a tough year for Australia's exports of coal for steel production and thermal coal, the Department of Industry and Science predicts various free trade agreements – particularly the one with China that removes the 3 per cent tariff on imports – will see the market rebound and continue to grow for a decade at least. The fact that China's coal-fired infrastructure is relatively new and has a predicted lifespan of forty to sixty years is often cited as the reason behind this optimism. 'China will be the coal giant for many years in the future,' according to the International Energy Agency's World Energy Outlook (2014).

While there is continuing debate about whether China's coal consumption has already peaked (International Energy Outlook 2015) or will peak and plateau in the 2020s as the Australian industry predicts, the Minerals Council suggests that demand from India will be such that Australian export earnings will continue to grow. Even if India meets its ambitious renewable energy targets, the Council appears confident that India's coal-fired capacity will have grown by 70 per cent by 2030. There is also increasing demand from Australia's close neighbours in Southeast Asia – the ten members of the Association of Southeast Asian Nations (ASEAN), which are turning to coal to drive their economic development over gas and oil (Minerals Council of Australia 2015). Malaysia is now one of the largest importers of coal in Asia, buying from Australia, Indonesia and South Africa (Quince 2015). Meanwhile, Japan was in 2015 the only member of the G7 looking to significantly increase coal-fired power generation, with 48 new projects in the pipeline. Overall, the world is predicted to use one billion more tonnes of coal a year in 2019 than it currently uses, and there is more investment in coal in the pipeline than for any other form of energy production.

Despite high-profile and targeted divestment and no-lending campaigns, banks in 2015 backed these growth predictions and continued to lend to the Australian coal and other fossil fuel industries. According to figures released by financial activists Market Forces, the top three lenders to Australian fossil fuels in 2015 were all Japanese financial groups – Sumitomo-Mitsui (A$2.87 billion), Mizuho (A$2.76 billion) and Mitsubishi UFJ (A$2.29 billion) – investing largely in gas but also maintaining Japan's historic interest in lending to the coal industry (Slezak 2016). Australian banks – the Commonwealth and ANZ – loaned the next largest amounts (A$1.75 billion and A$1.42 billion, respectively), followed by another Japanese lender, the Bank of Tokyo Mitsubishi (A$1.39 billion). Among the loans from Australian banks were eight with a combined value of A$4 billion for coal projects signed in 2015 (Slezak 2016).

The Australian industry shows few signs of willingly walking away from coal. Central to the industry's vision for the future are nine new mines in the massive Galilee Basin deposit in Queensland, 400 kilometres inland from Australia's northern east coast. If commissioned, the Carmichael mine, owned by the Adani Group, an Indian-based company, would produce 60 million tonnes of high-quality thermal coal a year (Adani 2014), a 'high-calorific value coal' that promises to deliver 'more energy with less coal, and fewer emissions' (Adani 2015). According to the company, 10,000 direct and indirect jobs would be delivered by its mine and associated rail and port projects. Taxes and royalties to the state of Queensland would amount to A$22 billion (Adani 2015). Also in the Galilee Basin, MacMines Austasia, now solely owned by the Meijin Energy Group, one of China's largest producers of coal and a major supplier of coal products to the United States, South Korea and Japan (through Mitsubishi), has sought approval to produce 38 million tonnes of coal a year from its China Stone Coal Project. The proponents have claimed that the project, with an expected mine life of fifty years, would create 3,900 jobs during the two-year construction phase and 3,400 positions once operational. Royalties would amount to A$5.9 billion over the mine's lifespan (Validakis 2015).

The pollutant

Coal is the source of two high-profile forms of atmospheric pollution – particles and greenhouse gases. In relation to the first, China has announced that it will burn less coal in order to control particle pollution (see Chapter 4), with public opinion surveys in China rating air pollution as the country's second biggest problem, behind corrupt officials. Although China still burns almost 4 billion tonnes of coal a year, estimates show a steady decline in coal consumption that is mostly affecting importers, including Australia (Grigg 25 September 2015). In the first eight months of 2015 alone, coal imports to China fell 31 per cent. Australia experienced a drop in thermal coal exports of 0.7 per cent in 2015, while exports of coking coal (used in steel production) rose modestly.

Particle pollution caused by coal has emerged as a political issue domestically in Australia. A 2013–2014 inventory of toxic substances found that coal was the leading source of particle pollution, with a doubling in coarse-particle pollution – PM10 – from coal mining in the previous five years. According to voluntary reporting, emissions of the more dangerous fine-particle pollution – PM2.5 – from the coal industry had increased by 52 per cent in the same period, compared to a general increase across all industries of 14 per cent (Cox and O'Brien 2015). The coal industry was responsible for 430,000 tonnes of coarse particle pollution in 2013–2014, or 47 per cent of the national total. Coal combustion also emits sulphur dioxide, nitrogen oxides and mercury, as well as other heavy metals (Union of Concerned Scientists n.d.). A much quoted fact and public relations problem for the coal industry is that more people die in Australia from air pollution than in car accidents (Cox and O'Brien 2015).

Coal is the major contributor of greenhouse gases, and it is estimated that coal from the Carmichael mine proposed for Queensland will produce more emissions in a single year than New York City (Taylor 2015). In terms of greenhouse gases, mining of coal directly releases methane, which the US Environmental Protection Agency suggests has a global warming potential twenty-three times higher than carbon dioxide. It estimates that coal mine methane contributes 8 to 10 per cent of human-made methane emissions worldwide. Electricity generation, of which 70 per cent relies on coal, contributes about 40 per cent of overall carbon dioxide emissions in Australia (Parliament of Australia 2010). While initiatives to lessen coal use and to introduce lower-polluting and more-efficient technologies are underway in Europe and the United States, increasing coal combustion in India and Southeast Asia is predicted to occur using inefficient subcritical technologies (International Energy Agency 2015). How this coexists with the 2015 COP21 agreement in Paris calling for the global increase in temperatures to be limited to 'well below' 2 degrees Celsius remains to be seen.

Despite incorporating 'clean air' measures such as 'carbon capture and storage' and 'high-efficiency super or ultra-supercritical plants' to address environmental concerns, the Australian coal industry's predictions of growth seem wildly optimistic in the face of the air pollution crisis facing China and global agreements on limiting emissions. Prices for coal are low. According to the International Energy Agency's (IEA) midterm report, prices of imported coal in Europe fell below US$50 a tonne in December 2015 – levels not seen in a decade. The IEA predicted that persistent oversupply and shrinking imports in China and elsewhere meant that coal prices would remain under pressure until 2020. Coal might still form the basis for a significant proportion of global energy production, yet its reputation has never been lower, with increasing pressure on governments, banks and corporations to divest from the industry and to shift to cleaner sources of power. The future of coal, it would seem, sits somewhere between the total industry collapse hoped for by climate change and other environmental activists and the rosy picture of continuing prosperity and even growth painted by the coal industry.

The Great Barrier Reef

Nowhere better illustrates the uncertainty and risk embedded within the coal debate than the Great Barrier Reef. Down the railway tracks from the massive coal deposits of inland Queensland, the reef is a vital producer of income and livelihoods for Australia – although economically it pales next to coal in terms of its overall contribution to the Australian economy. Described as one of the natural wonders of the world – 'a global nature superstar' – the reef is Australia's premier tourist destination, attracting more than A$5.7 billion to the Australian economy in 2012 and creating employment of almost 69,000 full-time equivalent workers (Deloitte Access Economics 2013). Stretching for 2,500 kilometres along the Queensland coast, the Great Barrier Reef is the world's largest coral reef ecosystem, and was listed by the World Heritage Committee in 1981 for its range of outstanding values, including being 'probably the richest area in terms of faunal diversity in the world' (UNESCO n.d.). Its scientific credentials are exceptional with a list of marine creatures that includes 600 types of soft and hard corals, more than 100 species of jellyfish, 3,000 varieties of molluscs, 500 species of worms, 1,625 types of fish, 133 varieties of sharks and rays and more than 30 species of whales and dolphins (Foxwell-Norton and Lester In press). The Great Barrier Reef Marine Park covers 344,400 square kilometres and contains 3,000 coral reefs, 600 continental islands, 300 coral cays and 150 inshore mangrove islands.

Culturally, the reef is part of Australia's national identity, with Australians defining themselves as coastal dwellers 'living on the edge' (Drew 1994). It is a site of historical and contemporary indigenous cultural heritage, retaining significance to Australia's Aboriginal people, especially those seventy Aboriginal and Torres Strait Islander Traditional Owner groups whose traditional lands border and include the reef (GBRMPA n.d. b), and it is the location of a brutal colonial history, invasion and frontier encounters, also part of the Australian identity (McCalman 2013). The reef supports numerous regional towns, communities and businesses reliant on tourism and fishing, and thus the continued status of the reef as a holiday destination. The impact of tourism-related activities, from outer reef snorkelling and diving visits to sunscreen use, is monitored for its potential threat to the reef's health.

However, it is the impacts of pesticide and soil run-off from coastal strip agriculture, coral bleaching, and changes to sea temperature and carbon dioxide levels associated with climate change that put the reef at most risk. Its management authority, the Great Barrier Reef Marine Park Authority (GBRMPA), warns of the impact of extreme weather events. The reef has long coped with cyclones and floods, but recent extreme weather events like those that occurred in 2010–2011 have caused unusual levels of damage. Tropical cyclones can cause extensive damage to individual corals and to the structure of the reef. According to the Authority, approximately 34 per cent of all coral mortality between 1995 and 2009 was caused by storm damage. Cyclones such as the Category 5 Yasi that hit in 2011 can have impacts that affect large areas for decades, if not centuries (GBRMPA n.d. a). Flood waters running into the shallow reef lagoon can also form, according to the Authority, reduced-salinity plumes laden with nutrients, sediments and agricultural chemicals such as fertilisers and pesticides, which stress and kill some of the reef's animals and plants, while encouraging productivity in others. Either way, the reef's ecosystem is disrupted. At best, the Great Barrier Reef is now described by its management authority as an 'Icon under Pressure' (Lloyd 13 August 2014).

Various Australian governments' eagerness to 'cut green tape' came very close in 2015 to confirming a new status for the reef with the UNESCO World Heritage Committee – 'in danger'. Of particular concern was the proposal that coal from the mines in the Galilee Basin would be transported to massively expanded shipping facilities at Abbot Point, on the central Queensland coast, where large-scale dredging would allow ships transporting the coal to make their way through the reef. How and to where the 3 million cubic metres of dredge spoil would be removed – the initial proposal to dispose of spoils within marine park boundaries was replaced by a proposal to dump on nearby wetlands valued by local indigenous communities – has caused continuing controversy. The new state government, while quickly reassuring investors of its commitment to the coal industry, has since proposed a second land-based site.

While this conflict over the reef is less surprising when viewed within the context of Australia's 'extreme' environmental politics this century, outlined in Chapter 2, it also illustrates the complexity of interactions that occur within environmental politics and communications – interactions between industry and government, media and political sources, science and activism. These are the 'switching points' of Manuel Castells' still emergent 'network society' (2004, 2009), where connectivity and power flow and clash to produce real outcomes for landscapes and the people who inhabit them.

Attributing responsibility for the reef

In this section, we examine mediated events that have occurred in relation to the Great Barrier Reef within the context of these transnational environmental, industrial and political pressures. Our focus is to identify and analyse discourses of responsibility: where and how they appear and, when possible, with what aims and to what effect. As we noted in Chapter 1, recent empirical research and theorising (see, e.g., Olausson 2009; Cerutti 2010; Jamieson 2010; Robertson 2010; Szerszynski 2010) have identified the attribution of responsibility as a key moment within a public sphere's discursive struggle over environmental harm, and the negotiation and distribution of justice more generally (Sen 2009: 337). Here, the struggle to contain spectacle is keenly fought and visible, and it is therefore vital that these critical moments within discourse are revealed. In attempting to do this, we apply two methods to consider media reporting of the issues facing the Great Barrier Reef – first, a qualitative analysis of media texts and their broader political contexts, outlined below, before applying a computer-assisted technique for text analysis. We end by highlighting our findings and with methodological reflections.

Qualitative analysis of responsibility attribution

Our approach focuses on claims-makers, changing media practices and technologies, and decision makers, analytically connecting media content with the social conditions and material culture of its production, use and flow (Appadurai 2008 [1990]) and identifying 'modes of symbiosis' (Morley 2009) between different media platforms. Following and analysing political messages and events as they move through media texts, phrases are identified in which 'responsibility' is attributed in relation to the Great Barrier Reef, alongside the political and media spheres in which the attribution is located. This analysis is cross-referenced and supplemented with interviews in Australia and Japan with environmental campaigners, government and industry representatives (including corporate and social responsibility officers, diplomats and corporate communications specialists), and journalists and other media producers.

Both the spectacular nature of the Great Barrier Reef and the stresses it is under frame media texts that attribute responsibility across various institutional, political and geographic arenas. Writing in August 2014 in the UK edition of the *Guardian* newspaper, for example, high-profile Australian scientist and environmental campaigner Tim Flannery attempts to assign rights and responsibility to distant publics:

> If the Carmichael coal mine is a global story, and the Great Barrier Reef a global asset, then the issue should not be left to Australia alone to decide. The citizens of the world deserve a say on whether their children should have the opportunity to see the wonder that is the reef. Opportunities to do this abound. Petitioning national governments to put climate change on the agenda

of the G20 summit, to be held in Australia in November this year, is one. Pushing governments to play a constructive role at the 2015 climate negotiations in Paris is another, as is letting the Australian government know directly that everybody has a stake in the reef, and that it needs to act to secure its future. The Great Barrier Reef does not have to die in a greenhouse disaster like the one that devastated the world's oceans 55 million years ago. But if we don't act decisively, and soon, to stem our greenhouse gas emissions, it will.

(Flannery 2014)

Flannery draws attention to the global and transnational elements of the case, defines the means for influencing international decision-making bodies, and by invoking the concept of an 'everybody', 'citizens of the world' and a global 'we', suggests the existence of a legitimate and potentially efficacious transnational public sphere (Fraser 2014). He is also assigning responsibility to a global 'we': 'if we don't act incisively, and soon'.

Greenpeace clearly spoke to the 'distant' when it warned that 'any dumping of dredge spoil on the World Heritage–listed reef will be an "international embarrassment" and akin to "dumping rubbish in the Grand Canyon"' (Petersen 2 February 2014). It further invoked the spectacular when it produced an advertisement that accused the Australian government of killing Nemo – in a blender no less. As reported by the *Daily Mail*:

> The super-cuteness of Nemo, the beloved clownfish made famous in Pixar's delightful film *Finding Nemo*, is being used to highlight what Greenpeace says is a potential environmental disaster on Australia's Great Barrier Reef. Greenpeace Australia Pacific has released a controversial advertisement which features a clownfish stuck swimming in a blender as part of its campaign to stop what it claims is a 'monstrous new mine' in Queensland, which will require a shipping terminal in the World Heritage listed Great Barrier Reef. The 30 second video, which was uploaded on YouTube on Tuesday, has since gained more than 29,000 likes.
>
> (Lee 2014)

Such appeals manifest across a range of local, national and international forums. Legal and governance structures are key spheres for drawing attention to the spectacular while publicly attributing responsibility, particularly given the well-established relationship between these institutional arenas and journalistic reporting practices. By January 2015, court cases against Adani and its Carmichael mine were underway in Australia. One was brought by the local Queensland Mackay Conservation Group, which claimed that the impact of greenhouse gas emissions on the reef had not adequately been 'taken into account' when the mine was approved (Chang 2015). In some reporting of this case, however, 'consequence' was expressed in terms of impact on the coal industry, and 'responsibility' placed on the conservation group for disrupting the industry and the federal government in approving the mine. A second case was brought by the Conservation Action Trust, an Indian environmental group, which was reported as being the first such challenge in Australia mounted by overseas activists. According to the *Guardian*:

> Debi Goenka, an executive trustee of the CAT, said: 'The coal from Carmichael, when burnt in India, threatens the health and livelihoods of poor, rural people in India. These people can't afford the electricity that will be generated – all they'll get will be damage to their health and the air, water, land and natural resource base on which their survival depends.'
>
> (Milman 9 October 2014)

Adani Mining's head Jeyakumar Janakaraj reportedly responded by claiming that activists were using lies in their anti-mining campaigns: 'I don't think they can sleep at night because they are using falsehoods' (McCarthy 2014). He drew on established corporate and social responsibility–type discourses of responsibility to restrict activist claims when he said: 'We are doing what is right. We are responsible, we are changing the lives of millions' (McCarthy 2014).

The struggle over the spectacular shifted into the political arena in November 2014 when US President Barack Obama made an official visit to Australia. In a speech at the University of Queensland, Obama told the audience the 'incredible natural glory of the Great Barrier Reef is threatened.' He located responsibility for the reef with the nation-state, and responsibility for climate change on nation-states collectively. While calling for a 'leapfrogging' of coal in developing countries, he also specifically queried the management of the reef and claimed the right of his daughters and their children to see the reef in fifty years' time. Australia's mismanagement meant they, too, were among the affected, he inferred. Both the Queensland and federal governments responded angrily. Claiming there 'was an issue' with the president's speech, the Australian foreign minister, Julie Bishop, said: 'We are demonstrating world's-best practice in working with the World Heritage Committee to ensure that the Great Barrier Reef is preserved for generations to come. . . . I think President Obama might have overlooked that aspect of our commitment' (Shanahan 2014).

Secondary appeals to consumers to alter their buying habits provide another sphere for the struggle to contain the spectacular and responsibility. 'Fight for the Reef' is a campaign jointly established by World Wildlife Fund (WWF)-Australia and the Australian Marine Conservation Society (fightforthereef.org. au). In April 2014, it achieved substantial publicity by winning the support of iconic US-founded ice cream company Ben & Jerry's, now owned by global retail giant Unilever. Under a campaign banner of 'Reef Scoop Tour', the company encouraged customers to 'Scoop Ice Cream, Not the Reef':

> We'll be travelling across our fair land, scooping out free ice cream and raising awareness of how the Reef is at serious risk from intensive dredging, mega ports and shipping highways, and encouraging Australians to join us.
>
> (Ben & Jerry's n.d.; see also Unilever n.d.)

Like Tim Flannery, WWF-Australia's chief executive officer, Dermot O'Gorman, invoked the notion of global shared concerns and responsibility when he described Ben & Jerry's involvement as reflecting the concern of people around the world about how the reef is being managed. Ben & Jerry's' tour is a timely reminder that the world expects the Queensland and Australian governments to 'lift their game' (*Brisbane Times* 2014).

In response, the Queensland government suggested Australians boycott Ben & Jerry's ice cream and referred the company to the Australia Competition and Consumer Commission. As in earlier examples of government and corporate responses, the government's reaction prioritised notions of 'truth' and 'fairness' as more important manifestations of 'responsible' behaviour. This, for example, was the response from the Queensland environment minister:

> Ben & Jerry's can campaign on whatever issue they like but as a company they have an obligation to tell Australians the whole truth and nothing but the truth. . . . Australia has strict laws to protect consumers against misleading and deceptive behaviour. These mistruths could cost jobs and development in regional Queensland. It's irresponsible behaviour from a company that should know better.
>
> (Vogler 2014)

Similarly, the Brisbane's *Courier Mail* stated:

> Ben and Jerry's ice cream has been hauled over the coals by the Queensland Government for supporting WWF's 'propaganda' to save the Reef campaign. Environment Minister Andrew Powell wants Australians to boycott the American company, saying it has damaged the reputation of the Reef and jeopardised jobs and tourism dollars. 'Another company has signed up to the campaign of lies and deceit that's been propagated by WWF,' Mr Powell said. 'The only people taking a scoop out of the reef is Ben and Jerry's and Unilever. If you understand the facts, you'd want to be boycotting Ben and Jerry's.'
>
> (Agius 2014)

The irony of the government's suggestion of a boycott of Ben & Jerry's was not lost on Queensland researchers Kerrie Foxwell-Norton and Marcus Lane (2014), who pointed out that meanwhile the federal Australian government had proposed legislative change to Section 45DD of the Australian Consumer and Competition Act removing exemptions for environmental and consumer campaigns so activists could no longer implement secondary boycotts as a protest strategy. As Foxwell-Norton and Lane write: 'Perhaps the Queensland Government missed the memo' (2014).

The principal site for the discursive battle over the reef has been UNESCO's World Heritage Committee, and specifically meetings in Doha in June 2014 and

Bonn in June 2015. While the Australian and Tasmanian governments 'accepted the umpire's decision' in relation to the 'humiliating' rejection by the World Heritage Committee of their attempt at Doha to delist 74,000 hectares of Tasmanian forests (ABC News 24 June 2014), it was reported that Australia's Department of Foreign Affairs had established a dedicated taskforce to ensure that the reef was not listed as 'in danger' by the United Nations (UN) (Milman 12 December 2014) when it next met in Bonn. Officials and ministers were dispatched around the world to lobby key countries over the issue, and international journalists and key decision makers were invited to Australia to visit the reef themselves. Australian ministers also raised the issue with member countries of UNESCO's World Heritage Committee on an opportunistic basis (Milman 12 December 2014). For the *Australian* newspaper, lobbying of the World Heritage Committee indicated the existence of 'deep international hostilities' over protection of the reef. Under the heading 'Reef rift exposed as campaign goes global', it reported:

> The federal government has banned dumping in Great Barrier Reef Marine Park waters and the Queensland government has promised to extend the ban to the remaining World Heritage boundaries that lie within state jurisdiction. The federal government is unlikely to be able to appease green groups, however. The government and resource groups say the true motive of the global campaign to protect the reef is to end coalmining, an issue that also lies at the heart of the UN's response to climate change. Greenpeace listed three concerns with the plan considered a key document in the UNESCO deliberations: it says it still allows coalmining, is silent on climate change and fails to address cumulative effects on the reef.
>
> (Lloyd 23 March 2015)

Nevertheless, the attempts to avoid responsibility for the reef's deterioration appeared unlikely to succeed if these reported comments from a member of the Portuguese delegation can be taken as representative:

> The major cause for the reef degradation is not only a consequence of extreme weather conditions and climate change as Australian Government documents seem to imply, but also due to human causes and interference. . . . We are concerned that not only Canberra is handing over environmental approval powers to the Queensland State Government on a matter of such high national and international relevance, but also other measures that have been taken that can deteriorate the health of the reef even more.
>
> (Sturmer 2014)

After the change of state government in Queensland in early 2015, it was reported that 'tough new regulations' to tackle the amount of pollution flowing onto the Great Barrier Reef would be considered, with the state's first ever 'reef minister' vowing to strengthen protections to avoid the ecosystem being listed as 'in danger' by the UN (Milman 18 February 2015). Meanwhile, the new government's decision to again move the dredge spoils dumping site was described by journalists as a 'symbolic change' and an indication of continued support for the development of the massive coal deposit. Premier Annastacia Palaszczuk was reported as saying her government 'sends a clear message: we can protect the Great Barrier Reef, and we can foster economic development and create jobs' (Lloyd 12 March 2015). The new government, however, was still attempting to shift responsibility, with journalists reporting that a government department was examining claims that Adani's 'chequered environmental and legal history' was grounds to revoke its status as a 'suitable operator' for Australia's largest coal mine. The department was reported to be considering an Environmental Justice Australia report that questioned how Adani Mining continued to pass its 'character check' in Queensland given the alleged role of related companies in 'serious legal violations and extensive environmental harm in India' (Robertson 25 February 2015).

Semantic analysis of responsibility attribution

In the following, we present and analyse the results of our application of automatic semantic analytical tools to media reporting of environmental issues facing the Great Barrier Reef. Semantic analysis is a particular type of natural language annotation – apart from part-of-speech tagging, syntactic parsing and other types of language-specific morphological analysis tools – which allows researchers to identify repetitions and associations within and across large sets of media texts. The key component of semantic analysis is a comprehensive semantic terminology containing a large variety of lexical groups, categories and classes – hierarchically structured and automatically mapped onto the raw media materials – to assist with textual and discourse analysis. Belonging to the family of natural language processing tools, semantic analysis has been widely adopted in the study of specialised genres and discourses such as product reviews and commercial promotion materials. While other natural language processing techniques can assist with the analysis of 'objective information' of texts like grammatical and syntactic structures, semantic analysis is considered useful in the extraction of patterns that underlie 'subjective information' such as attitudes and perspectives towards specific products, events and social phenomena.

Given that the primary purpose of semantic analytical systems is to extract information from large-scale databases regarding specific discourse features of environmental media and news reporting, they focus on general and abstract terms that reflect judgemental, evaluative and emotional language expressing. These include:

- Evaluation (good and bad, true and false, accuracy and appropriateness; authenticity)
- Importance (noticeability and markedness)

- States and processes (contentment; trepidation; apprehension and confidence)
- Personal traits (sensibleness and absurdity; strength and weakness)
- Relationship (obligation and necessity; competition and rivalry; power, authority and influence; permission and authorisation; help and hindrance)
- Psychological actions, states and processes (reasoning modes; belief and scepticism; knowledge, perception and retrospection; level of expectation; mental practices, procedures, resources and techniques; conceptual objects like ideas and concepts; level of interest, energy and boredom; desire and aspiration; effort and resolution; and intelligence and ability)

The semantic analysis tool we have used in this study of media reporting on the Great Barrier Reef and to ask how responsibility is attributed within these texts was developed by the Centre for Computer Corpus Research on Language of the University of Lancaster, UK. It is known as the UCREL Semantic Analysis System (USAS) (http://ucrel.lancs.ac.uk/usas/). The development of the version of the Lancaster semantic analysis tool deployed here relies on McArthur's *Longman Lexicon of Contemporary English* (McArthur 1992). USAS has a multi-tier structure which covers twenty-one major discourse fields and domains labelled alphabetically. Within each field, subdivisions are provided based on the semantic properties of terms and expressions classified in each major domain. In corpus discourse analysis, semantic properties refer to the inherent semantic correlation between different words and expressions. For example, the current version of the Lancaster semantic annotator highlights fifteen types of words and expressions which can be grouped into different categories of specific discourse functions (see Table 3.1). Words such as 'appropriate', 'disagree', 'inappropriate', 'suit', 'relevant' and 'unsuitable' which indicate the (lack of) appropriateness, suitability and aptness are marked by the code A1.2 The identification of such words can be done by retrieving such words from the database automatically. Such a 'search and retrieve' process is known as corpus semantic analysis.

Automatic corpus annotation tools like USAS claim to be instrumental in processing quantitative databases to extract useful textual information, and suggest that the statistical processing and modelling of media data are a prerequisite to conceptualising and developing theoretical constructs for empirical media studies. We have two aims in using this tool: one, to provide insights into our specific texts, especially the framing and editorial strategies devised for environmental media reporting purposes; and two, to consider the methodological continuity and disjuncture we identify between our two approaches.

As we argued in the book's opening chapter, the mining and extraction of subjective textual information – that is, language that is judgemental, evaluative and emotional – is useful in assessing the relationship between news media and public opinion and mobilisation. News language, apart from witnessing and reporting events of perceived social importance, is charged with ideologies, political interests, cultural predilections, personal motivations, values and attitudes (Fairclough 2013). It assumes the social role of informing and fostering a sense of responsibility among publics on social issues like climate change and environmental

Table 3.1 Sample semantic analysis categories

Semantic analysis category	Semantic category code	Explanation of analysis categories
Membership (S5–S8)	S5	Levels of association and affiliation (e.g. regroup, joint, bilateral, bond, community, together)
	S6	Levels of obligation and necessity (e.g. essential, mandatory, commitment, should, responsibilities, binding)
	S7.1	Power/authority/influence of administration/government (e.g. controlling, leading, convene, powerhouse)
	S7.4	Levels of permission, consent and authorisation (e.g. ban, allow, approve, ratify)
	S8	Levels of help/hindrance (e.g. barrier, obstacle, opposition, cooperation, promote, support, back, assist)
Cognition (X2.1–X2.6)	X2.1	Reasoning/thinking, levels of belief and scepticism
		(e.g. believe, formulate, view, attitude)
	X2.2	Levels of knowledge/perception and retrospection (e.g. knowledge, experience, awareness)
	X2.4	Investigation and examination (e.g. monitor, research)
	X2.5	Levels of understanding/comprehension (e.g. understanding, figure out)
	X2.6	Levels of expectation (e.g. expected, hope, due)
Interest and Involvement (X5.1–X9.2)	X5.1	Levels of attention (e.g. ignore, focus, concentrate, highlight)
	X5.2	Levels of interest (e.g. reluctant, grudgingly, energy, actively, keenly, interested, proactively)
	X6	(Lack of) decision
		(e.g. unresolved, concluded, finalised, finalising)
	X7	Levels of desire and aspiration (e.g. scheduled, aimed, target, ambitious, willingness, voluntary, plan)
	X8	Levels of efforts and resolution
		(e.g. efforts, trying, strive, done their best)
	X9.2	Level of success/failure (failed, succeeded, win-win, accomplished, taking off, solving)

Source: http://ucrel.lancs.ac.uk/usas/

protection (Dryzek 2013). An effective way to fulfil this role is to develop an affective news language style with the potential to promote attitude change and engage the general public in taking action to protect the natural environment. This specific case study on Australia explores potential advantages furnished by our methodological approach, with a view to developing useful empirical lines of research for environmental media reporting in distinct cultural systems.

To match the qualitative analysis based on recent environmental news reporting on the Great Barrier Reef, the news database compiled contains articles published in Australia between 2011 and 2015 which share the 'Great Barrier Reef' and 'pollution' as the two keywords in the textual content. It was discovered using the Environmental Health News database, a US-based foundation and reader-funded news service that both produces news and distributes news published in searchable databases from around the world (environmentalhealthnews.org). While not claiming total coverage nor to be free of editorial selection processes, this database was chosen for its extensive coverage of international environmental issues and comprehensive distribution network.

Following our keyword searches, the total size of our database was slightly over 500,000 words. The distribution of the news sources studied is shown in Figure 3.1, and include APN News Service, Australia ABC News, Australian Associated Press, *Australian Financial Review, Brisbane Courier Mail, Brisbane Times, Business Spectator, Gladstone Observer, Melbourne Age, National Times,* Science Network Western Australia, Sydney's *Daily Telegraph, Sydney Morning Herald, The Conversation* and *West Australian.*

As the size of data sets grows – an increasingly salient trend in many disciplines, including media studies – the computer-assisted approach to data analysis is deemed to play an instrumental role in advancing our understanding of new social and cultural events and phenomena in the media. The use of USAS identified a set of important text-internal features of the Australian media reporting on the conservation and debates over industrial development in areas close to

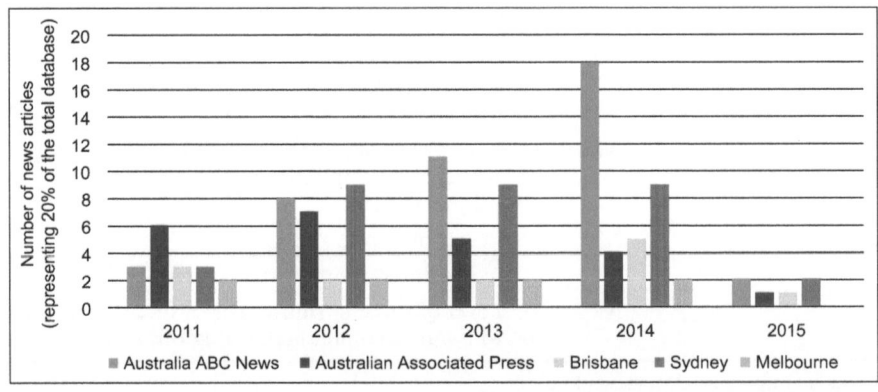

Figure 3.1 Structure of the Australian news data set used (in running tokens)

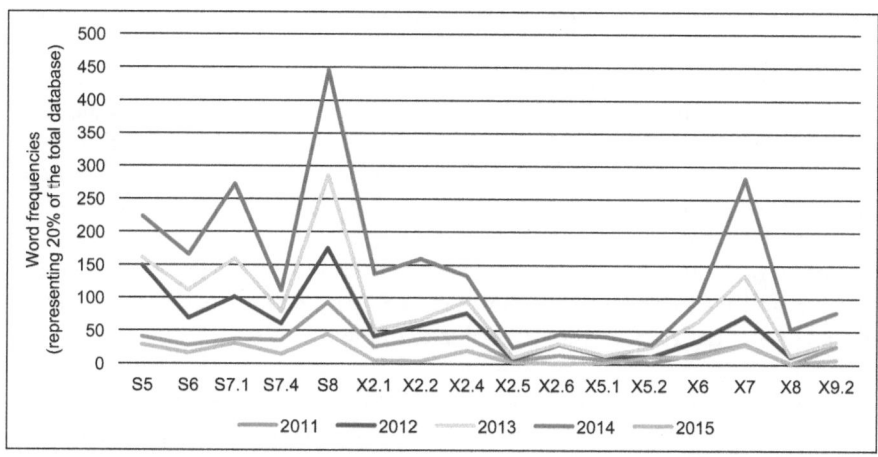

Figure 3.2 Great Barrier Reef reporting in Australia from 2011 to 2015

the reef since 2011. These features, underscored by a number of specific word groups, point to three key dimensions of the reporting on the reef: (1) Membership, (2) Interest and Involvement and (3) Cognition.

From these figures we are able to identify and extract some underlying patterns in the Australian media reporting on the reef at a cross-state level. A number of interesting textual patterns and features of the Australian media were examined. Overall, there are three peaks in Figure 3.2 suggesting the focus of the reporting on the Great Barrier Reef in Australia over the five-year period under investigation:

High-frequency words (Membership and Interest and Involvement)

- **X7**: levels of desire and aspiration (words such as 'scheduled', 'target', 'ambitious', 'willingness', 'plan')
- **S7.1**: influence of administration/government (represented by words such as 'controlling', 'leading', 'convene', etc.)
- **S8**: levels of help/hindrance (words such as 'barrier', 'obstacle', 'opposition', 'cooperation', 'support', 'assist')

The semantic category ***Membership*** includes five interrelated sublexical classes: (1) level of association and affiliation; (2) level of obligation or necessity; (3) influence of administration or government; (4) level of permission, consent and authorisation; and (5) level of help or hindrance to specific plans and/or actions. The semantic category ***Interest and Involvement*** is divided into six pertinent lexical classes: (1) levels of attention; (2) levels of interest; (3) (lack of) decision; (4) levels of desire and aspiration; (5) efforts and resolution; and, lastly, (6) judgement over the success or failure of specific actions.

The highlight of word groups from **Membership** and **Interest and Involvement** indicate that the Australian reporting on the environmental status of the Great Barrier Reef has its unique focus and features; that is, a clear and strong emphasis on multisectoral interaction, effective partnership building and adequate policy intervention to tackle social and research problems at a large scale such as the environmental impact of development on the Great Barrier Reef.

With regards to the prevalence of words from the semantic category of **Interest and Involvement** – that is, **X7** (desire and aspiration) – the corpus finding seems to show strong pressure within media discourse urging action: our close reading suggests that it is government and industrial sectors being urged to make practical and well-targeted plans and take concrete actions to combat existing and any potential risks to the reefs. For example:

> The Greens said the breakdown, that could cause 'significant environmental damage,' was another reason for the federal government to press pause *on plans to* (X7) increase shipping massively for fossil fuel exports throughout the Great Barrier Reef.
>
> (Whyte 2012)

> The stoush appears to raise doubts about *planned reforms* (X7) aimed at streamlining environmental approval of projects by shifting more responsibility to state governments.
>
> (Wroe 2012)

In the semantic analysis of the data set, the annotation category **Membership** is used to measure the levels and modes of collaboration between actors or stakeholders involved in the events or activities reported by the media. Under the **Membership** category, there are five subcategories of words which depict various aspects of partnership building and interaction among different societal sectors:

- First, S5: the subcategory of words indicating varying levels of association and affiliation. Typical words and expressions under the subcategory of association are 'regroup', 'joint', 'bilateral', 'bond', 'community' and 'together'.
- Second, S6: the subcategory of words describing the levels of obligation and necessity in tackling industrial pollution or other kinds of risks and threats posed by development activities near the reef. Typical words and expressions under the subcategory of responsibility are 'essential', 'mandatory', 'commitment', 'should', 'responsibilities' and 'binding'.
- Third, S7.1: the subcategory of words indicating the levels of power, authority and influence of administration and governments in tackling perceived risks to the reef. Typical words and expression under this category are 'controlling', 'leading', 'convene' and 'powerhouse'.
- Another subannotation category which sustains the corpus analysis membership building is S7.4, which illustrates levels of permission, consent and

authorisation of relevant industrial development proposals, activities and policies. These include words such as 'ban', 'allow', 'approve', 'disapprove', 'ratify', 'permit', etc.

* Lastly, words such as 'obstacle', 'opposition', 'cooperation', 'promote', 'support', 'back' and 'assist' are collectively grouped under the subcategory of S8. This is associated with the varying levels of help, support or hindrance from different societal sectors to the preservation of natural heritage sites such as the Great Barrier Reef.

A close observation of the textual patterns in the annotated corpus data points shed light on some interesting features of the reporting of the conservation of the reef in Australian newspapers. There are three coding categories which are high-frequency word groups in the reef reporting: S5, S7.1 and S8. Within the semantic analysis category of *Membership*, high-frequency words indicate prioritised aspects of partership or membership development around tackling environmental threats to the reef. The co-occurance of high-frequency word groups of *S7.1* and *S8* seems to suggest that effective policymaking and political intervention has provided the focus of much of the media debate on responsibility attribution around perceived risks and harms and actual damages caused to the Great Barrier Reef. For example:

there are three key factors that will determine if reefs can adapt: reducing of local stressors such as over-fishing, pollution and habitat destruction; expanding current *management* (S7.1) strategies such as marine protection zones, ecosystem-based *management* (S7.1) and water quality issues; and slowing climate change by aggressive reduction of CO_2 emissions.

(Cooper 2011)

That's why Labor wants to *convene* (S7.1) a high-level strategic group, including the primary industries community, to see the best way to reduce pollution run-off into the reef. We need our agricultural industry. Labor is proud to support our agricultural industry, but we also need to save the Great Barrier Reef.

(Smail 2015)

The World Wildlife Fund (WWF) says it is likely the World Heritage Committee will give Australia another year to strengthen its policies to *protect* (S8) the Great Barrier Reef off Queensland.

(Poyhonen *et al.* 2013)

The state government is to fund more than 30 research and *support* (S8) projects to help farmers from Mackay to Cooktown cut chemical run-off and soil erosion flowing into the ocean and killing coral.

(Twomey 2011)

These frequency-based textual features suggest that the issue of the reef is embedded in an especially salient political discourse, highlighting the importance of the influence exerted by the government and authorities (S7.1) in fostering social collaboration and joint efforts (S5) to effectively tackle the hindrance and difficulties (S8) faced by the conservation and sustainable development and use of the reef.

Lastly, as the corpus analysis shows, a unique feature of the Australian media reporting on the Great Barrier Reef is that the overall emphasis across this specific time period is given to words (*Cognition X2.4*), indicating the importance attached to investigation and examination (represented by words such as 'monitor', 'research') as opposed to words indicating levels of belief and scepticism or understanding and comprehension. A useful exploratory tool is the perspective of *Cognition* (a specific category of the semantic analysis) that is further divided into dimensions of belief and scepticism, understanding and comprehension; experience and awareness; expectation and lastly, (lack of) the use of scientific and investigative methods. Here are some examples:

> In a review article published in today's Science, Pandolfi (Professor John Pandolfi, of the University of Queensland's School of Biological Sciences) says *latest research* shows climate change remains the greatest threat to the world's reefs.
>
> (Cooper 2011)

> Discussing the health of the reef, the Environment Minister Tony Burke, yesterday said the number of ships working in the Great Barrier Reef will be *closely monitored* to ensure the health of the marine eco-system.
>
> (Phillips 2012)

> The *research findings* give hope that, even though warming of the oceans is already occurring, coral that has previously withstood anomalously warm water events may do so again.
>
> (ECOS 2012)

Conclusion

Two main methodological considerations provide the basis of the use of semantic analysis in our study of the Australian media reporting of the ecological impact of environmental changes associated with economic development and policy-making. First, as different from population health (see chapters on Japan and China), the impact of pollution on natural resources such as the Great Barrier Reef cannot be easily established or measured by mortality and morbidity. The subject of our case study represents a complex and politically controversial topic which provides the focus of heated and evolving debates in Australia and internationally. The nature of the subject we investigate, the Great Barrier Reef at risk and the roles and responsibilities of global industries, NGOs and decision-making bodies, thus requires an exploratory and descriptive approach to the analysis of Australian environmental media discourse. With its extensive (in terms of degree, level and scale) lexical categories, semantic analysis has proved a useful tool, which enables us to systematically explore the diversity and subtlety of the reporting and framing of the Great Barrier Reef under the continuing global pressure of coal development.

Second, as discussed earlier in this chapter, the looming environmental crisis which threatens the Great Barrier Reef points to the lack of effective and consistent policymaking over the issue. Multisector cooperation and partnership development on environment protection provides a much-needed yet largely underexplored social and research issue in Australia. Insights into the relationship between multisectoral collaboration and policymaking help fill a critical gap in the current study of environmental issues. This consideration has motivated us to use semantic analysis in our case study to explore the level of alignment or divergence among societal sectors, stakeholders and interest groups in Australia with regard to the perception and attribution of responsibilities associated with the increasing pollution and irreversible harms caused to the Great Barrier Reef. This research aim was achieved in the corpus analysis by examining the patterns of the distribution of words belonging to two semantic analysis categories, that is, *Membership* and *Interest and Involvement.*

On 1 July 2015, the UNESCO World Heritage Committee ruled against listing the Great Barrier Reef as 'in danger'. The decision followed lobbying and last-minute actions and promises made by the Australian and Queensland governments to ban dredging spoil dumping and cut pollution run-off by 80 per cent within a decade. Australia's environment minister was quoted as saying that the country had 'clearly heard the concerns of the world heritage committee' and implemented all its recommendations (Robertson 2 July 2015). The *Guardian* reported:

> The environment minister, who led a vigorous diplomatic lobbying effort to avoid an adverse listing following concerns raised by UNESCO last year, said some green groups had campaigned with 'spectacular lack of success' for the reef to be listed in danger.
>
> Greenpeace told delegates in Bonn that Australia's continued support for new coal mines in Queensland meant that there would be 'more dredging, thousands more coal ships through the reef and a dangerous amount of new coal being burnt.'
>
> Conservation groups have attempted to bring international attention to the difficulties faced by the reef. Greenpeace funded advertisements on London's Underground network telling commuters that this is their 'last chance to visit' the reef.
>
> (Robertson 2 July 2015)

According to news reports, the World Heritage Committee's decision was strongly influenced by very recent actions to curb the number of ports and to ban marine dumping of dredge spoil, but 'the elephant in the room' was still the Galilee coal basin, and the inevitable damage it would cause to the World Heritage area. Australia's plan to protect the reef was also largely silent on climate change, 'even though the Commonwealth's own Great Barrier Reef report card has identified climate change as the single largest threat to the reef' (Milman 2 July 2015).

Here, we see many of the important themes drawn out by our analyses at play. Both the qualitative and quantitative analysis revealed critical words in the news discourse, especially those charged with strong semantic meanings in establishing the patterns and modes of responsibility attribution around the environmental degradation of the Great Barrier Reef over recent years. For example, words in the qualitative analysis include 'symbolic change', 'suitable operator', 'serious violations', 'extensive harm', 'deep international hostilities', 'in danger', 'humiliating rejection', 'truth', 'fairness', 'responsible behaviour', 'pretty healthy', 'overlooked our commitment', 'consequence', 'monstrous new mine' and 'international embarrassment'. The quantitative analysis that retrieved keywords or, in statistical terms, high-frequency word groups, complemented the news discourse analysis. For instance, both approaches found that the scale of the impact of the changing environment, including development and climate change on the Great Barrier Reef, requires systematic research and effective policy and management intervention strategies. This is reflected in the high-frequency words in the news corpus indicating influence of administration/government (S7.1), levels of help and hindrance (S8) and investigation and examination (X2.4). This is supported by the qualitative analysis of recent news articles extracted from the database.

In a wider methodological context, there is systematic alignment between the corpus approach we used in this chapter with the quantitative corpus analysis that will be introduced in the two chapters on Japan; that is, the annotation category of *Membership* in USAS is methodologically connected with the coding schemes developed for the study of Japanese media reporting on air pollution and environmental innovation. For example, the five subcategories under *Membership* represent five important types of actions taken by multiple societal sectors, stakeholders and groups of interest in the development of the Australian discourse around the protection and conservation of emblematic national resources.

Our analysis of Australian environmental reporting has focused on the increasingly pressing problem of the deterioration and conservation of the Great Barrier Reef and its relationship to the country's historically embedded and still resonant connection to coal and reliance on global trade and tourism. In line with one of the main aims of this book, the case study is not meant to be an exhaustive investigation of news media reporting of threats to the Great Barrier Reef, but rather a contribution to knowledge and debate, and to methodological advances in empirical environmental media analysis. It demonstrates and illuminates how the deployment of automatic and exploratory data coding schemes such as USAS can assist with the extraction of useful patterns in quantitative media data sets, helping to further reveal the discourses identified qualitatively. It also shows us that the task we have set ourselves here is a potentially fruitful one; to access and distil large data sets and historically rich contexts in such a way that we and policymakers can recognise and expose the many subtle ways responsibility and blame can be distributed across complex communications, trade and political networks.

References

ABC News (24 June 2014) 'UNESCO rejects Coalition's bid to delist Tasmanian World Heritage forest'. Accessed on 14 August 2014 from www.abc.net.au/news/2014–06–24/unesco-rejects-bid-to-delist-world-heritage-forest/5538946

Adani (2014) *Carmichael Coal Mine and Rail Project*. Accessed on 14 August 2015 from www.adanimining.com/Australia-Carmichael-coal

Adani (15 December 2015) 'Land Court of Queensland recommends approval of Adani's Carmichael mine'. *Media Release*. Accessed on 9 February 2016 from www.adaniaustralia.com/media/media-releases

Agius, Kym (28 April 2014) 'Queensland government urges boycott of Ben and Jerry's ice cream over WWF propaganda'. *The Courier Mail*. Accessed on 23 March 2015 from www.couriermail.com.au/travel/australia/queensland-government-urges-boycott-of-ben-jerrys-ice-cream-over-wwf-propaganda-on-great-barrier-reef/story-fnjjv0r9–1226898567920

Appadurai, Arjun (2008) [1990] 'Disjuncture and difference in the global cultural economy', in J.X. Inda and R. Rosaldo (eds) *The Anthropology of Globalization: A Reader* (2nd edition). Oxford, UK: Blackwell Publishers, pp. 47–65.

Asafu-Adjaye, John (2009) 'Coal and the Australian economy', in Peter Knights and Michael Hood (eds) *Coal and the Commonwealth: The Greatness of an Australian Resource*, pp. 46–57. Accessed on 26 February 2016 from www.crcmining.com.au/wp-content/uploads/2013/05/Coal-and-the-Commonwealth_web.pdf

Australian Bureau of Statistics (2012) 'History of coal mining'. *Year Book Australia*. Accessed on 26 February 2016 from www.abs.gov.au/ausstats/abs@.nsf/Previousproducts/1301.0Feature%20Article1271910?opendocument&tabname=Summary&prodno=1301.0&issue=1910&num=&view=

Australian Government (n.d.) *National Heritage Places – Coal Mines Historic Site*. Accessed on 26 February 2016 from www.environment.gov.au/heritage/places/national/coal-mines

Ben and Jerry's (n.d.) *Scoop Ice Cream, Not the Reef*. Accessed on 14 August 2014 from www.benandjerry.com.au/flavours/reef-scoop-tour

Brisbane Times (14 August 2014) 'Ben and Jerry's ice cream hurting reef: Qld gov.' Accessed on 14 August 2014 from www.brisbanetimes.com.au/queensland/ben-and-jerrys-ice-cream-hurting-reef-qld-govt-20140429–37eg7.html#ixzz39ao4XvRH

Castells, Manuel (2004) *The Power of Identity* (2nd edition). Oxford, UK: Blackwell Publishers.

Castells, Manuel (2009) *Communication Power*. Oxford, UK: Oxford University Press.

Cerutti, Furio (2010) 'Defining risk, motivating responsibility and rethinking global warming'. *Science and Engineering Ethics* 16: 489–499.

Chang, Charis (16 January 2015) 'The court case that could choke mining in Australia'. *News.com.au*. Accessed on 23 March 2015 from www.news.com.au/technology/environment/the-court-case-that-could-choke-mining-in-australia/story-fnjwvztl-1227186867553

Colley, Peter (1997) 'Investment practices in Australian coal: The practice and profit of quasi-integration in the Australia-Japan coal trade'. *Energy Policy* 25(12): 1013–1025.

Cooper, Dani (22 July 2011) 'Rate of change key to reef survival'. *ABC Science*. Accessed on 18 July 2016 from www.abc.net.au/science/articles/2011/07/22/3273983.htm

Cox, Lisa and Natalie O'Brien (2 April 2015) 'Coal the biggest contributor to toxic air pollution study'. *Sydney Morning Herald*. Accessed on 26 February 2016 from www.smh.com.au/federal-politics/political-news/coal-the-biggest-contributor-to-toxic-air-pollution-study-20150401–1mcwbt.html#ixzz40JFelAGo

Deloitte Access Economics (2013) *Economic Contribution of the Great Barrier Reef*. Townsville: Great Barrier Reef Marine Park Authority.

Diamond, Marion (2009) 'Coal in Australian history', in Peter Knights and Michael Hood (eds) *Coal and the Commonwealth: The Greatness of an Australian Resource*, pp. 23–45. Accessed on 26 February 2016 from www.crcmining.com.au/wp-content/uploads/2013/05/Coal-and-the-Commonwealth_web.pdf

Drew, Philip (1994) *The Coast Dwellers: Australians Living on the Edge*. Victoria: Penguin.

Dryzek, John S. (2013) *The Politics of the Earth: Environmental Discourses* (3rd edition). Oxford, UK: Oxford University Press.

ECOS (10 April 2012) 'Could corals survive a warmer, more acidic ocean?'. Accessed on 18 July 2016 from www.ecosmagazine.com/print/EC12250.htm

Fairclough, Norman (2013) *Critical Discourse Analysis: The Critical Study of Language* (2nd edition). Abingdon, UK: Routledge.

Flannery, Tim (1 August 2014) 'The Great Barrier Reef and the coal mine that could kill it.' *The Guardian*. Accessed on 8 August 2014 from www.theguardian.com/environment/2014/aug/01/-sp-great-barrier-reef-and-coal-mine-could-kill-it

Foxwell-Norton, Kerrie and Marcus Lane (6 May 2014) 'Ben and Jerry's reef campaign shows that green groups are vital for democracy'. *The Conversation*. Accessed on 14 August 2014 from http://theconversation.com/ben-and-jerrys-reef-campaign-shows-that-green-groups-are-vital-for-democracy-26310

Foxwell-Norton, Kerrie and Libby Lester (In press) 'Saving the Great Barrier Reef from disaster: Media then and now'. *Media, Culture and Society*.

Fraser, Nancy (2014) *Transnationalizing the Public Sphere*. Cambridge, UK: Polity Press.

Great Barrier Reef Marine Park Authority (GBRMPA) (n.d. a) *Impact of Extreme Weather on Coral Reefs*. Accessed on 23 March 2015 from www.gbrmpa.gov.au/managing-the-reef/threats-to-the-reef/extreme-weather/ecosystem-impacts/impact-on-coral-reefs

Great Barrier Reef Marine Park Authority (GBRMPA) (n.d. b) *Traditional Owners of the Great Barrier Reef*. Accessed on 18 July 2016 from www.gbrmpa.gov.au/our-partners/traditional owners/traditional-owners-of-the-great-barrier-reef

Grigg, Angus (25 September 2015) 'China's "war on pollution" will hit coal hard'. *Financial Review*. Accessed on 26 February 2016 from www.afr.com/news/policy/climate/chinas-war-on-pollution-will-hit-coal-hard-20150925-gjv1k1#ixzz40JNdDTKB

Hood, Michael (2009) 'Introduction', in Peter Knights and Michael Hood (eds) *Coal and the Commonwealth: The Greatness of an Australian Resource*, pp. 8–22. Accessed on 26 February 2016 from www.crcmining.com.au/wp-content/uploads/2013/05/Coal-and-the-Commonwealth_web.pdf

International Energy Agency (2014) *World Energy Outlook 2014.* Accessed on 26 February 2016 from www.iea.org/newsroomandevents/pressreleases/2014/november/signs-of-stress-must-not-be-ignored-iea-warns-in-its-new-world-energy-outlook.html

International Energy Agency (2015) *Coal: Medium-Term Market Report 2015.* Accessed on 26 February 2016 from www.iea.org/Textbase/npsum/MTCMR2015SUM.pdf

Jamieson, Dale (2010) 'Climate change, responsibility and justice'. *Science and Engineering Ethics* 16: 431–445.

Lee, Sally (24 July 2014) 'They killed Nemo! New Greenpeace ad shows world's most beloved fish trapped in a BLENDER in grim environmental protest'. *Daily Mail.* Accessed on 23 March 2015 from www.dailymail.co.uk/news/article-2703691/New-Greenpeace-ad-shows-Nemo-trapped-blender-protest.html

Lloyd, Graham (13 August 2014) 'Outlook for Reef getting worse.' *The Australian.* Accessed on 8 August 2014 from www.theaustralian.com.au/national-affairs/climate/great-barrier-reef-outlook-getting-worse/story-e6frg6xf-1227022279428?nk=0af3f2c1626417b4037c07c802bdd1d9

Lloyd, Graham (12 March 2015) 'Palaszczuk government shows its faith in coal'. *The Australian.* Accessed on 23 March 2015 from www.theaustralian.com.au/national-affairs/state-politics/palaszczuk-government-shows-its-faith-in-coal/story-e6frgczx-1227259092309?sv=8c1cd9273d9f08324fd3d7ed8881df1e

Lloyd, Graham (23 March 2015) 'World Heritage Commission asked to Declare Reef in danger'. *The Australian.* Accessed on 23 March 2015 from www.theaustralian.com.au/national-affairs/climate/world-heritage-commission-asked-to-declare-reef-in-danger/story-e6frg6xf-1227273641263

McArthur, Tom (1992) *Longman Lexicon of Contemporary English.* UK: Pearson P T R.

McCalman, Iain (2013) *The Reef: A Passionate History from Cook to Climate Change.* Melbourne, Australia: Viking/Penguin Books.

McCarthy, John (28 November 2014) 'Premier Campbell Newman goes on resources offensive, hitting back at criticism from broadcaster Alan Jones and US president Barack Obama'. *The Courier Mail.* Accessed on 23 March 2015 from www.couriermail.com.au/business/premier-campbell-newman-goes-on-resources-offensive-hitting-back-at-criticism-from-broadcaster-alan-jones-and-us-president-barack-obama/story-fnihsps3-1227138847545

Milman, Oliver (9 October 2014) 'Carmichael mine: Indian conservation group joins legal battle with Adani'. *The Guardian.* Accessed on 23 March 2015 from www.theguardian.com/environment/2014/oct/09/carmichael-mine-indian-conservation-group-joins-legal-battle-with-adani

Milman, Oliver (12 December 2014) 'Great Barrier Reef: Australia sends diplomats out to defend its actions.' *The Guardian.* Accessed on 23 March 2015 from www.theguardian.com/environment/2014/dec/11/great-barrier-reef-australia-sends-diplomats-defend-actions-un-in-danger-list

Milman, Oliver (18 February 2015) 'Great Barrier Reef polluters face tougher action under Queensland's new government'. *The Guardian.* Accessed on 23 March

2015 from www.theguardian.com/environment/2015/feb/18/great-barrier-reef-polluters-face-tougher-action-under-queenslands-new-government

Milman, Oliver (2 July 2015) 'Great Barrier Reef: Australia says UNESCO decisions shows it is a "world leader"'. *The Guardian*. Accessed on 4 March 2016 from www.theguardian.com/environment/2015/jul/02/great-barrier-reef-australia-says-unesco-decision-shows-it-is-a-world-leader

Minerals Council of Australia (n.d.) *Characteristics of the Australian Coal Industry*. Accessed on 26 February 2016 from www.minerals.org.au/resources/coal/characteristics_of_the_australian_coal_industry

Minerals Council of Australia (2015) *Coal: Hard Facts* (2nd edition). Accessed on 26 February 2016 from www.minerals.org.au/file_upload/files/publications/Coal_Hard_Facts_2nd_Edition_FINAL.pdf

Morley, David (2009) 'For a materialistic, non-media-centric media studies'. *Television & New Media* 10(1): 114–116.

Olausson, Ulrika (2009) 'Global warming-global responsibility: Media frames of collection action and scientific certainty'. *Public Understanding of Science* 18: 421–436.

Parliament of Australia (2010) *How Much Australia Emits*. Accessed on 26 February 2016 from www.aph.gov.au/About_Parliament/Parliamentary_Departments/Parliamentary_Library/Browse_by_Topic/ClimateChange/whyClimate/human/howMuch

Petersen, Freya (2 February 2014) 'Great Barrier Reef Marine Park Authority approves plan to dump Abbot Point spoil'. *ABC News*. Accessed on 15 July 2016 from www.abc.net.au/news/2014-01-31/abbot-point-spoil-dredging-approved/5227774

Phillips, Nicky (7 March 2012) 'Great Barrier Reef is at a crossroads, says UN mission'. *The Sydney Morning Herald*. Accessed on 18 July 2016 from www.smh.com.au/environment/conservation/great-barrier-reef-is-at-a-crossroads-says-un-mission-20120306-1uit8.html

Poyhonen, Natalie, David Chen, and Kirsty Nancarrow (17 June 2013) 'More time tipped for Great Barrier Reef protection policies, WWF says'. *ABC News*. Accessed on 18 July 2016 from www.abc.net.au/news/2013-06-17/more-time-tipped-for-barrier-reef-protection-policies/4757720

Quince, Annabelle (29 September 2015) 'What future for coal, a crucial part of Australian history?'. *ABC Rearvision*. Accessed on 26 February 2016 from www.abc.net.au/radionational/programs/rearvision/what-future-for-coal-a-crucial-part-of-australian-history/6810886

Robertson, Alexa (2010) *Mediated Cosmopolitanism: The World of Television News*. Cambridge, UK: Polity Press.

Robertson, Joshua (25 February 2015) 'Adani's fitness to run Queensland mine examined over environmental concerns'. *The Guardian*. Accessed on 23 March 2015 from www.theguardian.com/australia-news/2015/feb/25/adanis-fitness-to-run-queensland-mine-examined-over-environmental-concerns

Robertson, Joshua (2 July 2015) 'UNESCO spares Great Barrier Reef "in danger" listing but issues warning'. *The Guardian*. Accessed on 4 March 2016 from www.theguardian.com/environment/2015/jul/01/great-barrier-reef-spared-unesco-in-danger-listing-un

Sen, Amartya (2009) *The Idea of Justice*. Cambridge, MA: Harvard University Press.

Shanahan, Dennis (21 November 2014) 'Julie Bishop "understands" fury at Barack Obama climate swipe'. *The Australian*. Accessed on 23 March 2015 from www.

theaustralian.com.au/national-affairs/climate/julie-bishop-understands-fury-at-barack-obama-climate-swipe/story-e6frg6xf-1227130123949

Slezak, Michael (26 February 2016) 'Australia's biggest banks pump billions into fossil fuels despite climate pledges'. *The Guardian*. Accessed on 26 February 2016 from www.theguardian.com/environment/2016/feb/26/australias-biggest-banks-pump-billions-into-fossil-fuels-despite-climate-pledges

Smail, Stephanie (14 January 2015) 'QLD Labor makes election promise to slash pollution in the Barrier Reef'. *ABC PM*. Accessed on 18 July 2016 from www.abc.net.au/pm/content/2015/s4162324.htm

Sturmer, Jake (19 June 2014) 'UNESCO ruling: Decision on whether Great Barrier Reef as "in danger" deferred for a year'. *ABC News*. Accessed on 23 March 2015 from www.abc.net.au/news/2014–06–18/unesco-defers-decision-on-great-barrier-reef-danger-status/5530828

Szerszynski, Bronislaw (2010) 'Reading and writing the weather: Climate technics and the moment of responsibility'. *Theory, Culture & Society* 27(2–3): 9–30.

Taylor, Lenore (12 November 2015) 'Coal from Carmichael mine "will create more annual emissions than New York"'. *The Guardian*. Accessed on 26 February 2016 from www.theguardian.com/environment/2015/nov/12/coal-from-carmichael-mine-will-create-more-annual-emissions-than-new-york

Twomey, David (21 November 2011) 'Bid to slash toxic run-off victory for Great Barrier Reef'. *EcoNews*. Accessed on 18 July 2016 from http://econews.com.au/9265/bid-to-slash-toxic-run-off-victory-for-great-barrier-reef/

UNESCO (n.d.) 'Great Barrier Reef'. Accessed on 8 August 2014 from http://whc.unesco.org/en/list/154

Unilever (n.d.) *Ben and Jerry's Joins Fight for the Reef*. Accessed on 14 August 2014 from www.unilever.com/brands-in-action/detail/ben-and-jerrys-joins-the-Fight-for-the-Reef/389282/

Union of Concerned Scientists (n.d.) *Coal Power: Air Pollution*. Accessed on 26 February 2016 from www.ucsusa.org/clean_energy/coalvswind/c02c.html#.Vs_c5zYwj0c

Validakis, Vicky (28 July 2015) 'New 6.7 billion coal mine proposed for the Galilee Basin'. *Australian Mining*. Accessed on 26 February 2016 from www.australian-mining.com.au/news/new-$6–7-billion-coal-mine-proposed-for-the-galilee

Vogler, Sarah (1 May 2014). 'LNP refers ice cream company Ben and Jerry's to ACCC over Barrier Reef campaign'. *Courier Mail*. Accessed on 14 August 2014 from www.couriermail.com.au/business/lnp-refers-ice-cream-company-ben-and-jerrys-to-accc-over-barrier-reef-campaign/story-fnihsps3–1226901781884?nk=1e25e3f6f89d4ceb1647c2366385b9af

Whyte, Sarah (19 May 2012) 'Cargo ship risk to Barrier Reef'. *Sydney Morning Herald*. Accessed on 19 July 2016 from www.smh.com.au/environment/conservation/cargo-ship-risk-to-barrier-reef-20120519-1yx8z.html

Wroe, David (6 June 2012) 'Mine threat to Barrier Reef: PM'. *Sydney Morning Herald*. Accessed on 19 July 2016 from www.smh.com.au/federal-politics/political-news/mine-threat-to-barrier-reef-pm-20120605-1zu9s.html

4 Responsibility and the risks of air pollution in China

Media activism and the case of *Under the Dome*

In China the politics of risk and responsibility for pollution intersects with a range of crucial issues for China's political, economic and social development, such as corruption, human rights, urbanisation, legal reform, the relationship between the state and the media and public participation in the political process. Several factors make it particularly important to examine the case of China. First, China plays a key role in the transnational political economy that produces significant amounts of pollution both within China and across the region. The continuing economic development of China is a major driver of environmental change, and any shift in Chinese attitudes and/or policies towards pollution is likely to have international repercussions. Second, the problem of pollution is at the centre of the dynamic relationship between state and society in China, and the media play an important role in this relationship. China's 'green public sphere' (Yang and Calhoun 2007) has grown in response to serious environmental problems, despite the fact that maintaining control over China's media is crucial to the Chinese Communist Party's efforts to govern the country effectively and keep its monopoly on power. How does the media deal with issues of environmental risk and responsibility in a country where the authorities view public discourse as a matter of national security?

Investigating how the politics of risk and responsibility for air pollution are reflected in the Chinese media requires a slightly different approach than in the cases of Australia and Japan. The Chinese Communist Party's sophisticated system of media control means that accurate quantitative news data – particularly from electronic sources – are often hard to obtain because news items can be changed or deleted from the Internet after publication. Unless monitoring of news data occurs in real time it is difficult to accurately measure what has actually reached the eyes and ears of Chinese audiences, especially when the topics are politically sensitive. For this reason the chapter shifts the focus away from macro-level quantitative analysis of how environmental risks and responsibilities are represented in the media and instead examines a micro-level case study – the documentary film *Under the Dome* [*Qiongding zhi xia*] – using a combination of quantitative linguistic analysis and qualitative analysis of the political constraints and opportunities that Chinese journalists must negotiate if they wish to actively shape the national discourse on air pollution risks and responsibility.

When former journalist Chai Jing released her documentary on air pollution, *Under the Dome*, just over a week before China's legislature, the National People's Congress, was due to meet in March 2015, she generated an explosion of discussion and debate on the subject and thrust the media politics of risk and responsibility for pollution into the national – and international – spotlight. Chai's documentary represents a case of activist journalism, albeit a somewhat unusual one. Chai was not officially employed as a journalist at the time she released her film, and yet a significant part of its success was due to her ability to leverage the official contacts she had made when working at the state-run broadcaster, China Central Television (CCTV), as well as her prominent reputation for environmental investigative journalism. The presentation of much of the film was similar to that of an in-depth news report, with sit-down interviews with officials as well as on-the-ground footage in the style of an investigative feature.

Activist journalists such as Chai have played a vital role in contributing to public discourse about the risks of and responsibility for pollution in China. Chinese journalists have often been at the vanguard of the ongoing struggle to protect the environment at a time of rapid industrialisation. However, China's political system, in which the media's official responsibility is to reflect the ruling party's interests rather than act as an independent check on power, and the influence of its propaganda system (*xuanchuan xitong*), which is designed to place certain restrictions on public discourse, have forced Chinese journalists to create new strategies for influencing debate over risk and responsibility. By combining an interpretive analysis of how *Under the Dome* deals with risk and responsibility for air pollution in China with quantitative linguistic analysis of the documentary's contents, this chapter investigates how one media activist successfully navigated the prevailing political impediments and intervened in the national debate over the state of China's environment.

This chapter begins with an overview of the growth of environmental risk in China, particularly in the form of atmospheric pollution, and then discusses the emergence of a distinct form of Chinese environmental activism. The chapter then explains the role played by journalists in both heightening public awareness of environmental risks and attributing responsibility for the current crisis before moving on to examine the case study in detail.

The case demonstrates how journalists' environmental activism in authoritarian states such as China can take hybrid forms that make use of both transnational and local communication strategies to intervene in national debates over risk and responsibility. Although *Under the Dome* employed a slick Technology, Entertainment, Design (TED) talk–presentation style that has been popularised in the United States and often displayed the characteristics of the kind of 'soft journalism' (Lester and Hutchins 2012) that has appeared in many other countries, these transnational communication styles were underpinned by a number of strategies that were tailored to navigate the particular political hazards of China's media environment. The communication strategy employed in this particular case of activist journalism mirrors the problem-solving logic of Chinese transnational environmental non-governmental organisations (NGOs), which draws on international resources but adapts them to fit local conditions (Reese 2015). The case also supports previous claims that Chinese journalists are pragmatic in their approach to activism and often strategically ally with forces inside the state to achieve their policy goals (Repnikova 2015).

In certain respects, however, the strategy that the creator of *Under the Dome* used in this attempt to publicise air pollution risk and responsibility represents a 'deviant case study' that defies some – but not all – of the normal expectations scholars have about activist journalism in China. Deviant case studies can be valuable when they help to refine prior theories and thereby generate a more robust understanding of the problem they illustrate (Lijphart 1971: 692). While the case of *Under the Dome* involves many of the standard goals and strategies of Chinese investigative journalists, the timing of the documentary's release and the scale of the problem that it tackles do not match the normal pattern of activist journalism in China.

The user – and severely affected

China's dramatic economic growth and rise (or return) to a position as one of the most powerful states in the world has been one of the most compelling stories of the late twentieth and early twenty-first century. Average economic growth rates of around 10 per cent for three decades have transformed China into a global economic powerhouse. Although China's growth is slowing and significant domestic structural problems have yet to be overcome, the country remains one of the key drivers of the world economy.

China's opening up to the world and integration with global markets has made this economic rise possible. Not only has China's growth been primarily driven by exporting industries in the manufacturing sector but this boom has been fueled by the import and consumption of a vast amount of natural resources. During the Maoist era China emphasised self-reliance and engaged in little trade with countries outside the socialist camp. This all changed as China moved into the era of reform and opening in 1978. China is now the world's largest consumer of energy, and so despite also being the world's largest energy producer it has been forced to look to international markets to meet its domestic demand. China became a net importer of oil in 1993 (see Downs 2004) and coal in 2009 (U.S. Energy Information Administration 2015: 27) and now accounts for approximately half of total global consumption of coal (U.S. Energy Information Administration 2015: 26). China's demand for coal is so great that it not only produces more coal than any other country, but it is also the world's largest importer of the commodity (U.S. Energy Information Administration 2015: 1). Although the proportion of the total energy supply that is produced by coal is lower in China than in the European Union or the United States, and Chinese households use much less energy than their Western counterparts, the use of coal in industrial processes such as steel and cement production is much higher than in the West, and the industrial and construction sectors consume approximately three-quarters of the energy produced in China (International Energy Agency 2014: 23).

Although the Chinese state and many individuals have benefited greatly from this economic growth, it has come at a high cost in the form of degradation of the natural environment and deteriorating public health. Problems of resource overuse and scarcity have been compounded by the damaging effects of pollution. China's water is so highly contaminated that the majority of the country cannot safely drink the water that comes out of the tap. Water pollution is particularly serious because China has so little freshwater to begin with – only one-quarter of the global per capita average (Li 2010). Water shortages have been triggered by sudden pollution crises (Chen 2009: xxi) but it is possible that in the near future there will be shortages in some northern Chinese cities even without the occurrence of such major events.

Pollution from pesticides and heavy metals not only threatens the water supply but also affects food safety (Lu *et al.* 2015). There is also a high level of public concern over the risk of food contamination through negligence or deliberate human intervention. A number of food-safety scandals have highlighted the risks associated with basic consumption in China and the lack of effectiveness of regulations designed to ensure quality control. The most high-profile of these cases was the contamination of milk powder used in baby formula with melamine, which was exposed in 2008 and resulted in the deaths of six babies and serious health problems for around 300,000 children across the country (Tracy 2010). This public concern over contamination has meant that organic produce is highly valued and rumours have occasionally circulated that China's leaders are fed a diet based wholly on their own secure supply of organic meat and vegetables. The least well off in society are often the most at risk of harm as they are the least likely to be able to pay for higher-quality domestic or imported consumer products.

Air pollution is the most visible form of pollution risk in China and at times has reached extreme levels. Air pollution is the leading global cause of environmentally related deaths and led to approximately seven million total premature deaths worldwide in 2012 (United Nations Environment Programme 2014: 44). One study estimated that 1.6 million deaths in China per year, equivalent to 4,000 per day and 17 per cent of all Chinese deaths, were attributable to PM2.5 pollutants in the air (Rohde and Muller 2015). In October 2012, heavy smog in Beijing forced the authorities to cancel flights and close major highways due to poor visibility despite the Beijing Environmental Protection Bureau claiming the air was only 'slightly polluted' (Caixin 2012). In January 2013, the PM2.5 level recorded by the United States Embassy in Beijing reached a staggering 'beyond index' reading of 755 (the normal range of measurement is from 0–500, with levels between 301 and 500 labelled 'hazardous') (Wong 2013). Air pollution affects life expectancy, and although the elderly can be acutely affected by high levels of pollution, it is likely to be the current generation of younger people who will pay the highest price in the long term (see Guo *et al.* 2013). Exercising outside in polluted urban areas can be a serious health risk; wealthy international schools in China have even begun to invest in inflatable domes with built-in air filters so that their students can safely participate in sports or exercise in a quasi-outdoor environment (Wainwright 2014). In major cities in China advertisements for air filter masks are common, and many families will either purchase or aspire to purchase an air filtering system for their home in order to manage the health risks posed by air pollution.

Public contestation over pollution risk and responsibility

As China's pollution crisis has worsened, a 'green public sphere' has emerged in which non-governmental organisations (NGOs) play a primary role in generating critical environmental discourse (Yang and Calhoun 2007). China's green public sphere, according to Yang and Calhoun (2007: 212), 'fosters political debates and pluralistic views about environmental issues' and constitutes a space where an issue-specific public can engage in 'nonpartisan' political advocacy through a variety of media and organisational forms. The 'greenspeak' that is generated in this public sphere focuses on volunteerism, participation and the development of an NGO culture, but it can also be a way to indirectly criticise government policies (Yang and Calhoun 2007: 216).

The emergence of this green public sphere has been facilitated by a number of sociopolitical transformations in China, particularly in the 1990s, that have opened up greater space for actors such as journalists, scientists, policymakers and social organisations to coalesce around environmental issues. These factors, which have been listed in more detail in Chapter 2, include the commercialisa-tion of the media and the authorities' decision to allow the managed growth of civil society groups. Some of these developments, such as advances in informa-tion communications technology and the growth of online social networks, as well as the increasing level of transnational interpersonal and interorganisational interaction between journalists, NGO workers and policymakers, reflect broader global trends rather than domestic policy choices, although the authorities' deci-sion to shift towards a managed process of opening up to the outside world did initially make it possible for these developments to affect China. The 'greening of the state' (Ho 2001), which has involved the development of a system of environmental laws and regulations, as well as new environmental institutions within the state apparatus, has also facilitated the emergence of China's green public sphere as the state has adopted the language of environmental protection and non-state and state actors have formed new networks.

Environmental issues have been at the forefront of public participation in politics in China's reform era, and at times public contestation over pollution risk and responsibility has come in the form of significant protests. These can be particularly explosive in rural areas when anger over environmental risk combines with other local issues such as ethnic tension, the abuse of power by officials or concerns over corrupt relationships between officials and business owners. For example, since 2011 Inner Mongolia has been the site of protests over the loss of grazing land caused by pollution, which threatens both the livelihood and cultural heritage of nomadic herding communities. In 2015, one person died and fifty were arrested when a protest against a chemical processing zone by ethnic Mongols was suppressed by approximately 2,000 riot police (*South China Morning Post* 2015). Urban residents have also been willing to engage in public protest to pressure the authorities to take action on environmental issues. The 'not-in-my-backyard' (NIMBY) movement, which has appeared in many middle class communities around the world and is based on local residents' opposition to industrial development projects, such as chemical factories or other polluting businesses that they perceive to be harmful to themselves or their interests, can now also be seen in many Chinese cities (Johnson 2013, 2010; Lang and Xu 2013).

Dramatic and contentious protests are only one aspect of environmental activism in China, however, and one that is often in opposition to the more cooperative bargaining strategy employed by most environmental NGOs (Johnson 2010). By aligning their actions with state objectives such as promoting good governance and raising public environmental awareness, Chinese citizens and NGOs can engage in environmental activism that largely avoids the realm of public confrontation while still attempting to reshape understanding of risk and responsibility. Unlike the environmental activism of post-Soviet states, these forms of political engagement do not constitute a social movement in opposition to the state and instead involve an ongoing process of interaction and negotiation with the state and the Party (Ho and Edmonds 2007a). The state also benefits from citizen engagement in environmental issues, and some cities have actively encouraged residents to participate in community forums to shape environmental policy. For example, since 2010 the Shenzhen government has heavily promoted its goal of increasing public input into its environmental decision-making process (Wu 2013: 102). New development projects are officially expected to conduct public consultations with local residents over their impact, including their potential to affect the environment, although this does not always occur in practice. The phenomenon of 'government-organised NGOs' (GONGOs) also confounds simple divisions between the Chinese state and society. The All-China Environment Federation (ACEF), which filed a 30 million–yuan public interest lawsuit against a chemical company for its pollution (Liu 25 March 2015), is a GONGO whose two most senior figures are a former state councilor and a former vice minister of the State Environmental Protection Administration (SEPA).

Social media has been crucial to the organisation of environmental protests and other forms of environmental engagement such as NGO activism. At the same time, however, social media provides opportunities for individuals to inform themselves about environmental issues and to express their views in a public forum or to a network of friends and associates without actively taking part in a public consultation or protest or joining an NGO. It is difficult for users of social media to remain unaware of the scale of China's environmental pollution and its corresponding health risks when some of the most egregious cases of environmental damage are widely and rapidly shared through online networks.

Online expressions of black humour and cynicism are a common outlet for a Chinese public frustrated and overwhelmed by regular revelations of new forms of environmental risk. When Beijing's authorities have ordered the temporary closure of factories, halted construction work and only allowed cars on to the road on alternate days for major political or sporting events such as the 2008 Olympics, the sky in the capital has become remarkably blue and free of pollution. When the sky cleared for the 2014 Asia-Pacific Economic Cooperation (APEC) summit after the authorities imposed strict temporary restrictions the public reacted cynically, labelling the colour of the sky 'APEC blue' and using the phrase to refer to any fleeting phase or artificial beauty (Wainwright 2014). When Beijing's air pollution again reached extreme levels in late November 2015 netizens shared tourist-style photographs of polluted vistas with the outlines of smog-obscured landmarks such as the CCTV building or Forbidden City drawn by hand onto the grey images.

Environmental activism and the news media in China

Journalists make an important contribution to the governance of environmental risk in China. When environmental NGOs began to emerge in China in the 1990s these groups were often led by journalists or former journalists (Yang and Calhoun 2007: 221–220). China's first legal environmental NGO, which was established in 1991, and some other influential early groups, such as Global Village and Green Earth, were founded by individuals with journalism backgrounds (Zhan 2011:117–118). In addition to their role in the construction of activist networks, Chinese journalists also play a more conventional role in shaping public discourse about risk and responsibility in ways that encourage government officials and individuals to take action to help tackle environmental problems. Many media outlets participate in state-led environmental educational campaigns to raise public awareness of environmental risks. Individual activist journalists also uncover and draw attention to environmental risks through investigative reporting. Many such journalists view their target audience as not the mass public but rather the elites who have the power to effect meaningful change in state policy or practice (de Burgh and Zeng 2012: 1015). When major environmental crises occur, or when activist journalists engage in critical investigative reporting, the media apportion responsibility for environmental problems, both in terms of blame for causing them as well as duty to act in response to them. When a chemical spill in 2005 caused the contamination of an entire city's supply of drinking water the state-run media outlet CCTV blamed not only the company involved but also the environmental regulators who had failed to prevent the spill and criticised local officials for attempting to cover up the crisis (Tilt and Xiao 2010).

Chinese environmental journalists face a well-understood tension between the pressure to respond to the preferences of their paying audience and the requirement that they act as the voice of the Party on important political issues and shape public perceptions in a way that generates support for the Chinese

Communist Party (CCP) (see Zhao 1998). China's authoritarian state places significant constraints on journalists through its use of a range of tools, including a strict licencing regime for journalists, a regulatory system that requires all media outlets to secure sponsorship from an administrative unit in the state or Party, a pervasive propaganda apparatus that issues direct instructions to media outlets and influences key editorial appointments, ambiguous state secrets laws that harshly punish leaks of official information and well-developed journalistic norms that proscribe reporting on certain politically sensitive topics (see Brady 2008).

Political constraints such as these do not mean, however, that journalists must choose between directly opposing the authorities or unthinkingly conforming to Party dictates. Maria Repnikova (2015) has shown that the practice of investigative journalism in China challenges the common understanding of journalism in authoritarian political systems, which presents journalists as either acting as agents for democratic change by holding leaders to account for their actions or contributing to the resilience of the authoritarian system by providing the appearance of openness and tolerance while lacking any real influence. Refuting this dichotomy, she argues that Chinese journalists do play an important role in improving governance through accountability but that they view this in pragmatic terms and often strategically ally with forces inside the state to achieve their policy goals. In China's fragmented authoritarian system (Lieberthal 1992) where the law offers little defence against state power it is particularly beneficial for activist journalists to build alliances with sympathetic agencies or individuals within the state, not only to gain access to information, as is the case for journalists operating in liberal democratic states, but also to provide a measure of protection against retaliation by the targets of their campaigns. Although some media campaigns are clearly driven by government agendas and others take environmental discourse in directions that place pressure on officials to change their behaviour and may lead to tension between state actors and journalists, the reality for journalists is that campaigns that challenge the interests of certain state actors will often assist other actors within the state to achieve their goals.

In order to retain their voice in China's developing green public sphere and to shape discourse on pollution risk and responsibility it has been necessary for environmental journalists to adopt similar tactics to their counterparts working in Chinese environmental NGOs (ENGOs). Ho and Edmonds (2007a, 2007b) argue that Chinese environmentalists operate in a semi-authoritarian system that presents them with both constraints and opportunities, which produces 'embedded activism'. They show that although the authorities regard the involvement of foreign actors in Chinese NGOs as a potential threat to state security and strictly prohibit activists from organising in direct opposition to the Party or state, the ability of environmentalists to develop informal networks with those inside the Party or state who share similar goals is a major asset for activists attempting to effect policy reform. Chinese environmental activists also tend to enact a form of 'depoliticized politics' that avoids confrontation with the central authorities (Ho and Edmonds 2007b: 336). Similarly, Elizabeth Perry (2007, 2008) has argued that Chinese protests in general tend to take the form of an appeal to higher levels of government to enforce existing rules and that activists frame their demands using the language of the authorities. Like ENGOs, activist journalists engage in a form of 'negotiated symbiosis' (Ho and Edmonds 2007a) with the state. An acute awareness of the political constraints they face, combined with an entrepreneurial approach to coverage of environmental issues (Yang 2010), is needed if activist journalists are to navigate the Chinese authorities' complex system of media controls. In the next section this chapter examines how this manifests in practice through an in-depth case study of a prominent example of activist environmental journalism.

The case of *Under the Dome*

In late February and early March 2015 the issue of air pollution exploded into public consciousness in China like never before with the online release of the film *Under the Dome*. This documentary about air pollution was made and fronted by high-profile environmental journalist Chai Jing, who is a former anchor and investigative reporter for China's main state-run broadcaster, China Central Television (CCTV). In 2014 she resigned from her job at CCTV in order to care for her daughter, who was undergoing treatment for a benign tumour (Kuhn 2015), but also spent this period making *Under the Dome*. Chai claimed that the film cost around 1 million yuan to produce and that she used her own money from royalties for an earlier book to fund the project after refusing offers of support from various foundations in China (*China Digital Times* 2 March 2015). Chai was already widely known in China for her investigative reporting on pollution, particularly in her home province of Shanxi, a major centre for coal mining, and in 2007 she had been awarded the government's 'Green Chinese' prize for environmental protection (*China View* 2007). However, the release of *Under the Dome* dramatically increased her domestic and international profile.

The film itself screened in Beijing on Friday, 27 February 2015. It was released online the next day and immediately went viral. By Sunday afternoon it had attracted more than 25,000 comments on the video-sharing website Youku (Buckley 2015), and by Monday it had generated 280 million posts on Sina Weibo, China's version of Twitter, and been viewed more than 100 million times (Gardner 2015). The websites of two ENGOs mentioned in the film – Friends of Nature and the Institute of Public and Environmental Affairs – crashed on the Sunday due to overwhelming demand (Bannister 2015). In less than a week it attracted more than 300 million views online (Branigan 2015). The impact of the film attracted comparisons with Rachel Carson's groundbreaking book *Silent Spring* and with Al Gore's climate change film *An Inconvenient Truth*, while Chinese environmentalist Ma Jun called the documentary 'one of the most important pieces of environmental awareness building ever in China' (Ma 2015).

Under the Dome is produced in the style of a TED talk, with a casually dressed Chai Jing addressing a studio audience from a stage and speaking without notes. Throughout the presentation her speech is interspersed with large-screen graphics and short prerecorded film clips. The documentary's title references an American television show based on a Stephen King novel about a town that is suddenly covered by a giant dome. Chai reveals that she realised it was an appropriate metaphor for the air pollution issue when she saw her daughter banging on the windows in their apartment in her frustration at not being able to go outside on smoggy days.

The film begins with Chai Jing introducing the major smog that Beijing experienced in January 2013, her discovery around that time that she was pregnant and her subsequent worries as her newborn daughter underwent surgery for a tumour that had been discovered while she was still in the womb. The birth of her daughter made Chai Jing reassess her previous lack of concern over the health effects of air pollution (she had not previously worn a mask when outside) as she realised how vulnerable her baby was and that she was completely responsible for everything that her daughter ate, drank or breathed. She sets out to answer three questions in the film: what is smog, where does it come from and what do we do about it [*women zenme ban*]? The first question addresses the risks of air pollution and the latter two address the issue of responsibility.

Risk and responsibility for air pollution in Under the Dome

The film portrays the risk of air pollution in terms of its impact on human health in stark terms. In answering the question 'what is smog?' Chai explains what PM2.5 is, the fact that it contains fifteen different carcinogens and affects the lungs and heart and measures how much of it she encounters in one day of normal life. Her result of 305.91 micrograms per cubic metre (μm^3) is four times the Chinese health guidelines, which are themselves three times the WHO guidelines of 25 $\mu g/m^3$. The film shows children born with respiratory problems, and Chai points out the correlation between increases in PM2.5 levels and increases in death rates and the number of children and elderly people admitted to hospitals. She cites former Minister of Health Chen Zhu as saying that every year air pollution leads to the death of 500,000 people in China. In this section the film also attempts to puncture some of the myths that people often use to rationalise their acceptance of the risk of air pollution, such as the idea that if children are exposed to it from a young age their bodies will adapt to the toxic environment. She also admits that the news media has previously downplayed the risk of air pollution by describing heavy pollution in Beijing as 'fog'.

Rather than limit her analysis to one location or specific case of pollution, which we would expect if Chai were to follow the usual pattern of investigative reporting, instead she does not hesitate to link pollution to issues affecting different parts of the country. For example:

> But afterwards I came to know that in Beijing the PM2.5 pollution caused by the food and beverage industry accounts for almost 6 per cent; the food and beverage industry in Beijing represents 12 per cent of the whole country – such a big restaurant.

And:

> Among the ten global major sea ports, seven are now in China. Ocean freighters have caused so much pollution that in places around 400 metres off the coastline the impact of the pollution is equal to that produced by 500,000 large trucks. So 66 per cent of SO_2 in Shenzhen is from freighters.

At the same time, however, the film also raises some of the less tangible risks of air pollution and environmental degradation. One of the film's most power-ful scenes is a clip recorded when Chai was working in Shanxi in 2004 where a 6-year-old girl tells Chai that she has never seen a real star or a white cloud. Chai also reveals that during the APEC CEO Summit in Beijing in 2014, when the government clamped down on polluting industries and the skies suddenly cleared, her husband showed her a place in Beijing where his father had taken him ice skating, swimming and fishing when he was a boy. She compares the moment to a child looking at the last sweet, knowing that if they do not eat it the sweet will melt but also that if they do eat it there will be none left. Such intangible losses due to pollution are interwoven in the documentary with the more direct risks to human health.

Under the Dome discusses the responsibility for air pollution in a number of ways. Chai first explains the sources of air pollution and from those sources identifies the key contributors to China's smog crisis. She identifies coal, oil, biofuel, industry, agriculture, fertiliser and dust as sources but then shows that 60 per cent of China's PM2.5 is from burning coal and oil. She points out that China burns more coal than the rest of the world combined and that the last time one country was so dominant was the UK in 1860. Surprisingly the province where the greatest burning of coal in China occurs is not Shanxi, which is well known for coal mining, but Hebei, which is a major centre of steel production. Chai also explains that China is in an unusual position because it is industrialising in a much more compressed time period than the developed world and so it must deal with the problems of burning coal and the problems of burning oil at the same time, rather than deal with each problem in turn. In Beijing, where Chai lives, 800,000 new cars were added to the roads in 2012, and vehicle pollution is the biggest source of PM2.5.

After many interviews with government officials and academics, as well as footage of Chai's travels around China and overseas as she documents the sources of pollution, how authorities attempt to regulate it and the views of business owners and other individuals about the environment and current efforts to manage pollution, at around the one-hour point the film summarises the key reasons for China's air pollution crisis: overconsumption of oil and coal, consuming relatively low quality oil and coal and the failure to clean fuel or to adequately regulate emissions. She blames overconsumption not on individuals but on urbanisation and an overheated economy that funnel huge amounts of state capital into construction that far exceeds demand. Chai claims that if the development plans of China's major metropolises were an accurate reflection of

China's real population then China would be a country of 3.4 billion people. She quotes a Tsinghua professor who says that excessive urbanisation means that if China does not shift to new ways of allocating capital 'the pollution and traffic jams have only just begun.'

For the most part, however, Chai attributes primary responsibility for China's air pollution crisis to the country's regulatory environment. She lists the following specific problems: China has no requirement to wash coal prior to using it; there is no way to stop polluting companies; it is not clear who is responsible for vehicles that do not meet emissions regulations, and these vehicles cannot be recalled; there is no way to improve fuel standards (due to the oil industry's control of the regulatory process); and no one has the authority to actually inspect fuel products. As she explores these specific problems two key themes emerge: lack of regulatory enforcement and collusion between polluting businesses and powerful state interests.

Chai shows that China does not lack environmental laws but rather lacks effective enforcement. She interviews a researcher from the Chinese Academy of Science who states bluntly that if the existing environmental protection regulations were enforced this would reduce ash and sulphur dioxide emissions by 60 per cent and nitrogen oxide emissions by 35 per cent. Even when polluters are caught they often manage to avoid any repercussions. Regardless of how diligently individual Ministry of Environmental Protection (MEP) inspectors work to identify and report polluting factories they do not have the power to ensure that the polluters are punished. She shows a clip of inspectors stopping a truck on its way into Beijing because it does not have the required emissions-controlling device, despite displaying a green sticker indicating it complies with environmental regulations. The inspectors are unable to fine the driver because he is transporting vital supplies (eggs, milk and oil) into the city and so is exempt from the normal regulations.

Chai's approach goes beyond a simple appeal to the state for greater enforcement of existing laws and regulations, however, and actually attempts to analyse why the rules are not being properly enforced. She explains that despite the existence of consumer protection laws, since 2004 no vehicles have been recalled due to faulty environmental labels because – according to the relevant regulatory authority – they do not pose a direct safety threat to the public. Since laws on atmospheric pollution were introduced in 2002, factories producing illegal vehicles could be shut down and fined and their products can be confiscated, but Chai points out that this has never happened since then because the regulations do not specify exactly which department should take responsibility for enforcing this law. She does note that the Atmospheric Pollution Prevention Law will be revised later in the year and the current draft gives the MEP the authority to issue fines for or recall fraudulent vehicles, although this would not come into effect until September 2015 at the earliest.

The film also takes the significant step of blaming collusion between polluting businesses and powerful state interests for China's failure to solve its air pollution crisis. Chai shows that state subsidies for the steel industry are resulting in oversupply, and a senior MEP official in Hebei bluntly admits that the steel industry is too important to the economy of the province for unlicenced factories to be shut down. A major issue that she covers here – and the most politically sensitive part of the film – is the lack of competition in the energy sector. Major state players dominate the market and use national security to justify keeping private firms out while rejecting any rise in fuel standards that might threaten profits. Chai shows that a lack of competition is not only linked to corruption in the sector but also means that China's market for natural gas, which is a much cleaner source of energy, has remained relatively underdeveloped, with only 5 per cent of the country's energy coming from natural gas compared to the global average of 24 per cent.

Although Chai generally avoids singling out anyone for direct criticism, the person whom the film portrays in the least flattering light is Cao Xianghong, who is the head of the National Oil Fuel Standards Committee and former chief engineer at China's giant state-owned oil and gas company Sinopec. When Chai asks Cao why the process of setting fuel standards is controlled by the major corporations he argues that those outside the oil industry, including the MEP, do not understand the oil refining business and therefore should not be involved. When Chai questions him about why the costs of increasing the standards are not made public he makes the rather strange argument that since people already do not trust Sinopec it is better not to say anything because some people will seize on any information that is released and be excessively critical. He also cites the importance of national security considerations in the oil industry and claims that an interruption in the fuel supply due to higher regulatory standards could lead to social instability. When pressed as to why Sinopec, as a large and highly profitable state-owned enterprise (SOE), cannot take on more social responsibility for the environment he simply says that Sinopec is very 'fat', but it is all fat and no muscle. Although the word 'responsibility' is spoken only three times in the documentary, two of those come in Chai's conversation with Cao Xianghong when she asks about Sinopec's social responsibility.

The final part of the film covers the question of what can be done about smog. Here the focus shifts from identifying who or what is responsible for causing the problem to a discussion of who is responsible for solving the problem. The film again raises the issue of China's regulatory system and enforcement measures but Chai tempers her criticism with the claim that 'even the most powerful government in the world can't control pollution by itself' and spends the last ten minutes of the film explaining what kinds of actions individuals can take. She argues that the severe acute respiratory syndrome (SARS) crisis taught her that public participation requires information disclosure and discusses an app that individuals can use on their smartphone to identify and report polluting industries in their local area. Hebei has introduced legislation requiring key polluting industries to publicise their data and if they do not individuals can sue them. Although NGOs have previously not been allowed to file lawsuits because the regulations state that only 'relevant organisations' (*youguan zuzhi*) can do so, from January 2015 an organisation can sue if it has been carrying out environmental protection work for at least five years and has not broken the law. She also shows a video made in cooperation with Friends of Nature that explains what actions individuals can take if they are unable to join an environmental NGO, such as not driving for any trips less than 5 kilometres, calling a hotline to report diesel vehicles that are emitting black smoke or to report excessive dust from construction sites, cleaning the kitchen range hood at home and asking restaurants to install filters, boycotting the products of polluting manufacturers and participating in the public consultation process when environmental laws and regulations are being drafted.

Response to the film

The initial reaction to the film from government sources was generally positive. The new Minister for Environmental Protection Chen Jining praised the film, calling it a 'Silent Spring' moment for China (Gardner 2015), and sent Chai a text message to thank her for drawing attention to environmental problems (Ng 2015). At the opening of the National People's Congress (NPC) on the Thursday following the film's release, Premier Li Keqiang called pollution 'a blight on people's quality of life and a trouble that weighs on their hearts' and said that 'we must fight it with all our might' (Agence France-Presse 5 March 2015). At the premier's annual press conference, held at the close of the NPC session, Li was asked about *Under the Dome* by a journalist from the Huffington Post. Without specifically referring to the film he responded by emphasising that the government was determined and had made significant effort to tackle pollution but also admitted that the progress did not meet public expectations and claimed that the focus for the coming year would be on enforcement measures and implementing the new environmental protection law, which had been passed the previous year (Mufson 2015). On the Tuesday after the film came out the Ministry of Science and Technology released a draft of a five-year air pollution prevention and control policy. The website of the *People's Daily*, the main mouthpiece of the CCP, also published a lengthy interview with Chai about the film.

Markets in China also reacted strongly to the film. Shares in environmental companies surged on the Monday after the documentary was released, with several rising by the daily limit of 10 per cent (Shao 2015). At the same time, international interest in the film was high, with major newspapers around the world covering the story and posting links so that their readers could watch it themselves.

A few days after the film's release, however, the authorities began to issue instructions to remove it from the web and clamp down on online discussions of the topic. Leaked censorship orders revealed that there was concern amongst officials that the issue would overshadow the major political meetings of the NPC and the Chinese People's Political Consultative Committee (CPPCC) that were taking place in Beijing that week and that some people would use the documentary as an opportunity to attack the government (*China Digital*

Times 3 March 2015). In a press conference held at the end of the week the minister for environmental protection avoided mentioning the film (Wong and Buckley 2015) and Chai's interview on the *People's Daily* website was removed. Employees at the *Global Times*, a popular nationalist newspaper controlled by the People's Daily Group, said that the paper's editors had been forced to cancel plans to publish a series of articles and opinion columns with a range of perspectives – both positive and negative – on the film (Wong 2015).

Praise for Chai continued over the following months despite the online clampdown. In April 2015 *Time* named her one of the world's 100 most influential people and in early June Chai was given an environmental protection award by the Chinese environmental group Society of Entrepreneurs and Ecology (Liu 5 June 2015). In late 2015 *Foreign Policy* included her in its annual list of 100 leading global thinkers (*Foreign Policy* 2015).

Although not directly attributable to *Under the Dome*, throughout 2015 the government continued to publicly emphasise the steps it was taking to tackle pollution. The National Development and Reform Commission (NDRC) annual report for 2015, which was announced at the NPC meeting in the week following the film's release, pledged to 'attach greater importance to green, circular, and low-carbon development' (State Council 2015), and in mid-March regulators stated that they were aiming to have a new system of environmental taxes, which had been expected to take another two years to implement, in place by the end of 2015 (Kong 2015). At the end of March it was reported that the MEP was reviving the concept of 'green gross domestic product' (GGDP) and planned to begin reintroducing it in 2016–2017 (Li 2015). Around the same time the major environmental GONGO, the All-China Environment Federation (ACEF), filed a 30-million-yuan public interest lawsuit against a chemical company over air pollution (Liu 25 March 2015). The suit was made possible by the introduction of a new environmental protection law, which came into force at the beginning of 2015, allowing NGOs to sue polluting businesses in class action–style cases. In May 2015, the State Council announced that the importance of environmental factors in official performance assessments of cadres would be 'significantly increased' and that responsibility for any environmental damage that occurred under an official's watch would continue even after the official had left his or her post (Li 2015). By late 2015, the Beijing authorities had closed two coal-fired power stations and opened a gas-powered plant, and an energy official has stated that by the end of 2016 the city will no longer produce electricity from coal (McDonell 2015). In late November 2015, as Beijing again choked on hazardous levels of smog, Environment Minister Chen Jining announced that China had met its latest pollution reduction targets six months ahead of schedule (Xinhua 2015).

In the months following the film's release President Xi Jinping's anti-corruption campaign targeted individuals connected to Sinopec, the state-owned oil company that was heavily criticised in the documentary. In late April, the Central Commission for Discipline Inspection (CCDI), which is the CCP's main anti-corruption organisation, announced that Sinopec's second-most senior executive, General Manager Wang Tianpu, was suspected of breaking the law and violating party discipline (Guo 2015). In September, he was expelled from the CCP and the CCDI announced that he would face charges for taking bribes, extortion and other abuses of power (Blanchard and Chen 2015). In October, the governor of Fujian Province, Su Shulin, who is a former general manager of Sinopec, was placed under investigation for 'serious violations of discipline' (Associated Press 2015). These moves cannot be directly attributed to the influence of *Under the Dome*, however, as the focus on corruption in the energy sector began well prior to the film's release. In mid-2014 an investigation into former Politburo Standing Committee member Zhou Yongkang, who had deep connection to the oil industry, was announced; when Zhou was found guilty in April 2015 and given a life sentence the following June he became the most senior Chinese political figure ever to be convicted on corruption charges since the founding of the People's Republic (British Broadcasting Corporation 2015). It is possible that rather than the documentary affecting the authorities' choice of target, the ongoing investigation into corruption in the energy sector gave Chai greater confidence to criticise the major state-owned enterprises in the film.

Navigating the political constraints on media activism

The content of Chai's film paints an overwhelmingly negative portrait of the current condition of China's environment and at times is sharply critical of regulatory bodies and other state actors. In essence the documentary draws attention to the risks associated with air pollution, explains that regulatory failures are largely to blame and shows how individuals can take greater responsibility for solving the crisis by becoming more outspoken and active in reporting environmental problems. The overwhelming public response and the attention from China's top leaders indicates that *Under the Dome* was a highly successful intervention in China's national debate over air pollution risk and responsibility. To understand this success it is necessary to examine how Chai's production cleverly makes use of techniques that not only increase the effectiveness of her message but also assist in negotiating the political constraints on activist journalism in China.

At first glance Chai's presentation seems to resemble forms of environmental journalism and media activism that have previously appeared around the world. The documentary employed the human-interest narratives and personalised focus of the 'soft journalism' that Lester and Hutchins (2012) have observed in Australian environmental journalism, while its TED talk presentation style led to comparisons with Al Gore's famous climate change film *An Inconvenient Truth*. Although Chai skilfully employs the kinds of presentation techniques that could be used by environmental activists in many different countries, the analysis here shows that *Under the Dome* is very much a product of its political context.

To some extent Chai's documentary follows a pattern of activism that is fairly well established in China. Despite the clear political significance of *Under the Dome*, Chai's approach appears to reflect a desire to engage in a form of 'depoliticized politics' (Ho and Edmonds 2007b: 336) that avoids direct confrontation with the central authorities. Perhaps the most obvious strategy involves Chai's attempts to personalise the risks and consequences of air pollution. Although this is a very effective communication technique with which to build rapport with her audience, it also serves to depoliticise the message. In addition to beginning the film with a discussion of her concern for her daughter's health she talks about waking up every day to check the pollution level and planning her day around it. Chai describes her anti-pollution efforts as a personal battle and says that she does not want air pollution, and her constant concern for her daughter's health, to dominate her life. When she uses the film to challenge conventional thinking she admits that she once held incorrect views about the pollution problem. This makes her arguments seem more like a collaborative learning process than a lecture. After discussing how individuals can take action she shows a scene where she goes to a construction site – uncovered dirt blows into the air and Chai successfully gets them to cover it. There is also a scene where she prompts the restaurant at the bottom of the building where she lives to install an air filter. She reveals her family's decisions about car use and ends the film by discussing her daughter's health and talking about our responsibility to the next generation. She makes a direct link between protecting the environment and protecting her daughter. Throughout the film she presents herself as an ordinary person rather than a member of any organisation or movement.

Although Chai represents *Under the Dome* as an independent and highly personal project, it seems clear that the film had the support of key players in the state and in the official media. The film therefore resembles the 'embedded activism' (Ho and Edmonds 2007a, 2007b) often practiced by Chinese environmentalists, which involves the negotiation of strategic alliances between civil society and the state. The film was first broadcast on the website of the CCP's official newspaper *People's Daily*, and Chai herself talked about how government officials sent her a lot of information to help with her investigations. On the Tuesday after the film came out the Ministry of Science and Technology released a draft of a five-year air pollution prevention and control policy, which seems to indicate that policymakers were well prepared for the documentary's release. The fact that the film was released online one day after the environment minister Zhou Shengxian was transferred out of his post and a new minister appointed also seems more than just a coincidence.

Chai is careful not to directly criticise senior officials and instead uses central government statements or policies to support her position on what should be done about air pollution. Xi Jinping makes a brief appearance in the film when Chai plays a clip from his APEC speech where he says, 'I hope and believe that through hard work we can maintain our "APEC blue",' but the main senior government figure she relies on to support her argument is NDRC vice-chairman Xie Zhenhua. In addition to his position in the NDRC, which is the top government organ in charge of economic development, Xie is also China's chief climate change negotiator. Chai uses her interview with Xie to counter the argument that environmental protection will be too much of a drag on economic growth. Not only does Xie point out that environmental protection 'is an industry too' and cite employment figures in this area, he also says that China can learn from the experiences of London and Los Angeles and can even solve its problems faster than they did. Chai also refers to high-level official statements to support her points. She cites the government's agreement during the APEC summit that its carbon emissions would peak around 2030 and renewables would constitute 20 per cent of energy sources and also quotes China's 2014 national energy security strategy, which states that energy is a commodity, market mechanisms need to be built and the methods of supervision need to change.

Chai avoids interviewing environmental activists and instead uses the words of polluters or recalcitrant officials to reveal their attitudes and motivations in order to shed light on the pollution problem. In one revealing scene an official from the MEP has his identification documents snatched away by a manager of a polluting business, and when the official says that he has the obligation to carry out inspections he is met with the rejoinder 'you have the obligation but not the authority.' In another scene a business owner claims he would never carry out construction work without the relevant building permits or tax registration but when asked why he has proceeded without the relevant environmental permission he simply answers that he will eventually get around to obtaining it. Chai quotes another business owner who says the lack of enforcement means he must ignore environmental regulations if his business is to remain competitive. Rather than represent these officials or business owners as the 'enemy', which would involve a more confrontation approach to the problem, she attempts to reveal the motivation for their illegal behaviour.

The way that Chai presents foreign perspectives in the film appears to indicate that she is acutely aware of the sensitivity surrounding foreign involvement in Chinese civil society and foreign criticism of China's political system. One of the film's main rhetorical strategies is to compare China and Beijing to cities in the developed world that have suffered from similar problems with air pollution in the past, such as London, Los Angeles and Tokyo. The film even shows footage of people in London wearing face masks to combat the smog produced by burning coal. London and Los Angeles are both powerful comparisons because each suffered through an era of major air pollution before putting in place policies that have had a great degree of success in dealing with the problem. Although many people in China argue that it took London fifty years or more to clean up its smog, Chai shows that in fact pollution levels dropped by 80 per cent in the first ten years after new regulations were introduced, largely through a shift to natural gas and oil, and this did not slow down economic growth. She says directly that London's experience shows that China needs to leave the age of coal behind and move into the era of oil and natural gas. Controlling the pollution from cars is another focal point of Chai's comparisons with other major cities. She points out that despite Tokyo's level of car ownership being very high, fewer than 6 per cent of people in Tokyo drive to work, whereas 34 per cent of Beijing residents drive. She explains how cities such as London have reduced the number of cars in the city by increasing parking charges and shows that even though the number of cars in Los Angeles has tripled since the 1970s emissions have gone down by 75 per cent over the same period due to tough new regulations. In the United States, the Environmental Protection Agency (EPA) has genuine teeth to investigate and fine car factories that produce vehicles that do not meet emissions standards and to recall vehicles that do not meet standards. Chai also points out a case where two Chinese people were jailed and faced a very large fine in Los Angeles for vehicle labelling fraud.

As she presents these comparisons, however, Chai stresses that she is not arguing that foreigners are more virtuous than Chinese people. When the US authorities tried to introduce stricter emissions standards for vehicles the American auto industry also complained and tried to block the changes. Once foreign car companies said that they could comply with the stricter rules the US companies followed suit, although they lost half their market share in the process. She makes the point that human nature and corporations' narrow focus on profits are the same all over the world but this can be controlled if the law is enforced effectively. Chai's questions for her non-Chinese interview subjects tend to involve asking them to explain how environmental regulations work in their own city rather than asking them to comment on China's problems, thus avoiding the sensitivity surrounding foreign interference in Chinese affairs and reducing the potential for her to be accused of being overly sympathetic to foreign critics. After the film's release she refused an interview request from the British Broadcasting Corporation (BBC), who claimed that Chai and her production team were 'wary of the foreign media' (Hatton 2015). Despite the strong international and domestic interest in the film she only gave one interview, which was to the *People's Daily* (Kuhn 2015).

In some respects, however, *Under the Dome* represents a departure from the usual forms of activist environmental journalism in China. Repnikova (2015: 10) points out that during politically sensitive periods 'the space for investigations tends to tighten, and then expand again once the risks of political instability appear to subside.' To many people the film's release on Friday, 27 February 2015, only a few days before the start of the annual CPPCC and NPC sessions, which opened on 3 March and 5 March, respectively, indicated that Chai was making a direct attempt to influence the agenda of the two meetings. Normally at such times editors are under strict instructions from the propaganda authorities not to print or broadcast anything that might distract the public or hijack the political agenda and must instead produce stories that adhere closely to the official line. The way that the film tackles its subject matter is also rather unusual. Activist journalism in China tends to avoid systemic issues and instead focuses on small-scale governance failures at the local level with the aim of attracting the attention of the higher authorities (Repnikova 2015: 11–12). *Under the Dome*, in contrast, is relatively fearless in tackling large-scale systemic problems that prevent effective regulation of the environment, not only in Beijing but all over the country. Chai's focus on deep systemic problems, her wide-ranging coverage of many different locations and interconnected issues as well as her success in getting her film to the public at a highly sensitive time indicates that *Under the Dome* is an exception to the normal patterns of activist journalism in China.

Conclusion

This chapter has investigated how the politics of pollution risk and responsibility play out in the Chinese media through a case study of Chai Jing's air pollution documentary *Under the Dome*. The combination of China's unwritten yet very real restrictions on media reportage of sensitive issues and its broader constraints on political activity has generated a distinctly Chinese form of environmental activism. Activist journalists such as Chai are embedded in networks that include actors from both the state and civil society sectors. They employ the kinds of political communication techniques found in other countries but adapt them to fit local circumstances. When they are successful they can play a key role in heightening public awareness of environmental risks and attributing responsibility for China's current environmental crisis.

The case of *Under the Dome* serves both to confirm and to challenge previous assumptions about how activists use the media to shape public understanding of environmental issues in China. Many of the common features of 'embedded activism' are present in Chai's attempt to intervene in the national debate over air pollution risk and responsibility. Yet the case also demonstrates how much can be accomplished even within the limitations imposed by China's system of controls over public discourse. While the political constraints may be real, an astute and well-connected activist such as Chai can stretch them to a greater extent than had been previously assumed to be possible.

Chai's extensive experience as a CCTV journalist made her an expert in the 'political ritual' of Chinese television news (Chang and Ren 2016). The political savvy shown by Chai in deciding which techniques to employ and when to push back against the normal limitations that constrain journalistic activism in China played an important part in maximising the chances that the documentary would reshape the discourse surrounding air pollution risks and responsibility. Her ability to ally with partners in the Chinese state likely made it possible for her to stage a major media intervention during a time of heightened political sensitivity. Although in many ways *Under the Dome* appears to be a heroic individual effort to spark a national debate about air pollution risk and responsibility, the film actually highlights the fact that environmental activism in China is embedded in a web of networks and opportunity structures that encompasses both the media and the state.

References

Agence France-Presse (5 March 2015) 'China vows to fight pollution "with all our might"'. *The Guardian*. Accessed on 19 July from www.theguardian.com/environment/2015/mar/05/china-vows-to-fight-pollution-with-all-our-might

Associated Press (8 October 2015) 'Former head of China's Sinopec under investigation for corruption'. *The Guardian*. Accessed on 19 July from www.theguardian.com/world/2015/oct/08/former-head-of-chinas-sinopec-under-investigation-for-corruption

Bannister, Tom (2 March 2015) 'China's latest silent spring: FON and IPE's websites crash after Chai Jing's documentary goes viral'. *China Development Brief.* Accessed on 19 July from http://chinadevelopmentbrief.cn/articles/chinas-latest-silent-spring-fon-ipes-websites-crash-chai-jings-documentary-goes-viral/

Blanchard, Ben and Chen Aizhu (17 September 2015) 'Update 1-former Sinopec executive to be prosecuted for graft'. *Reuters.* Accessed on 19 July from www.reuters.com/article/2015/09/18/china-corruption-sinopec-corp-idUSL4 N11O1KA20150918

Brady, Anne-Marie (2008) *Marketing Dictatorship: Propaganda and Thought Work in Contemporary China.* Lanham, MD: Rowman and Littlefield.

Branigan, Tania (5 March 2015) 'Beijing authorities sanguine as pollution documentary takes China by storm'. *The Guardian.* Accessed on 19 July from www.theguardian.com/world/2015/mar/05/beijing-sanguine-pollution-documentary-china

British Broadcasting Corporation (BBC) (11 June 2015) 'China corruption: Life term for ex-security chief Zhou'. Accessed on 19 July from www.bbc.com/news/world-asia-china-33095453

Buckley, Chris (1 March 2015) 'Documentary on air pollution grips China'. *Sinosphere.* Accessed on 19 July from http://sinosphere.blogs.nytimes.com/2015/03/01/documentary-on-air-pollution-in-china-grips-a-nation/

Caixin (21 January 2012) 'Beijing releases PM 2.5 air quality readings'. *CaixinOnline.* Accessed on 19 July from http://english.caixin.com/2012–01–21/100350762.html

Chang, Jiang and Hailong Ren (2016) 'Television news as political ritual: Xinwen Lianbo and China's journalism reform within the party-state's orbit'. *Journal of Contemporary China* 25(97): 14–24.

Chen, Gang (2009) *Politics of China's Environmental Protection: Problems and Progress.* Singapore: World Scientific.

China Digital Times (2 March 2015) 'Translation: Interview with Chai Jing'. Accessed on 19 July from http://chinadigitaltimes.net/2015/03/translation-peoples-daily-interview-chai-jing/

China Digital Times (3 March 2015) 'Minitrue: Clamping down on "Under the Dome"'. Accessed on 19 July from http://chinadigitaltimes.net/2015/03/minitrue-clamping-dome/

China View (15 December 2007) 'Director Zhang Yimou wins "Green Chinese" award'. Accessed on 19 July from http://news.xinhuanet.com/english/2007–12/15/content_7252954.htm

de Burgh, Hugo and Rong Zeng (2012) 'Environment correspondents in China in their own words: Their perceptions of their role and the possible consequences of their journalism'. *Journalism* 13(8): 1004–1023.

Downs, Erica (2004) 'The Chinese energy security debate'. *The China Quarterly* 177: 21–41.

Foreign Policy (2015) *Chai Jing: For Making China's Silent Spring.* Accessed on 14 September 2016 from http://2015globalthinkers.foreignpolicy.com/#!stewards/detail/jing

Gardner, Daniel K. (18 March 2015) 'China's "silent spring" moment? Why "Under the Dome" found a ready audience in China'. *The New York Times.* Accessed on 19 July from www.nytimes.com/2015/03/19/opinion/why-under-the-dome-found-a-ready-audience-in-china.html

Guo, Aibing (28 April 2015) 'Sinopec no. 2 caught in graft probe as China crackdown widens'. *Bloomberg*. Accessed on 19 July from www.bloomberg.com/news/articles/2015–04–27/sinopec-no-2-faces-corruption-probe-as-china-crackdown-widens-i902nvcx

Guo, Yuming, Li Shanshan, Tian Zhaoxing, Pan Xiaochuan, Zhang Jinliang, Williams Gail *et al.* (2013) 'The burden of air pollution on years of life lost in Beijing, China, 2004–2008: Retrospective regression analysis of daily deaths'. *BMJ* 347(f7139). http://dx.doi.org/10.1136/bmj.f7139

Hatton, Celia (2 March 2015) 'Under the dome: The smog film taking China by storm'. *BBC News*. Accessed on 11 August 2016 from www.bbc.co.uk/news/blogs-china-blog-31689232

Ho, Peter (2001) 'Greening without conflict? Environmentalism, NGOs and civil society in China'. *Development and Change* 32(5): 893–921.

Ho, Peter and Richard Louis Edmonds (eds) (2007a) *China's Embedded Activism: Opportunities and Constraints of a Social Movement.* London: Routledge.

Ho, Peter and Richard Louis Edmonds (2007b) 'Perspectives of time and change: Rethinking embedded environmental activism in China'. *China Information* 21(2): 331–344.

International Energy Agency (2014) *Coal Medium-Term Market Report 2014: Market Analysis and Forecasts to 2019.* Paris: International Energy Agency.

Johnson, Thomas (2010) 'Environmentalism and NIMBYism in China: Promoting a rules-based approach to public participation'. *Environmental Politics* 19(3): 430–448.

Johnson, Thomas (2013) 'The health factor in anti-waste incinerator campaigns in Beijing and Guangzhou'. *The China Quarterly* 214: 356–375.

Kong, Lingyu (19 March 2015) 'Amid worsening pollution, gov't moves toward law on ecotaxes'. *Caixin*. Accessed on 19 July from http://english.caixin.com/2015–03–19/100792892.html

Kuhn, Anthony (4 March 2015) 'The anti-pollution documentary that's taken China by storm'. *NPR.org*. Accessed on 19 July from www.npr.org/sections/parallels/2015/03/04/390689033/the-anti-pollution-documentary-thats-taken-china-by-storm

Lang, Graeme and Ying Xu (2013) 'Anti-incinerator campaigns and the evolution of protest politics in China'. *Environmental Politics* 22(5): 832–848.

Lester, Libby and Brett Hutchins (2012) 'Soft journalism, politics and environmental risk: An Australian story'. *Journalism* 13(5): 654–667.

Li, Biao (31 March 2015) 'Lüse GDP yanjiu gezhe 11 nian hou chongqi, ming hou nian shidian "2.0 ban"' [Green GDP research returns after being shelved for 11 years, pilot "version 2.0" next year and year after]. *Meiri Jingji Xinwen*. Accessed on 19 July from www.nbd.com.cn/articles/2015–03–31/906606.html

Li, Jiao (2010) 'Water shortages loom as northern China's aquifers are sucked dry'. *Science* 328: 1462–1463.

Lieberthal, Kenneth (1992) 'The "fragmented authoritarianism" model and its limitations', in Kenneth Lieberthal and David M. Lampton (eds) *Bureaucracy, Politics, and Decision Making in Post-Mao China.* Berkeley, CA: University of California Press, pp. 1–30.

Lijphart, Arend (1971) 'Comparative politics and the comparative method'. *The American Political Science Review* 65(3): 682–693.

Liu, Qin (25 March 2015) 'China court to hear 30m yuan air pollution lawsuit'. *Chinadialogue*. Accessed on 19 July from www.chinadialogue.net/article/show/single/en/7790-China-NGO-files-3-m-yuan-lawsuit-against-polluting-firm

Liu, Qin (5 June 2015) 'China's widely-watched pollution documentary scoops major award'. *Chinadialogue*. Accessed on 19 July from www.chinadialogue.net/blog/7962-China-s-widely-watched-pollution-documentary-scoops-major-award/en

Lu, Yonglong, Shuai Song, Ruoshi Wang, Zhaoyang Liu, Jing Meng, Andrew J. Sweetman, Alan Jenkins, Robert C. Ferrier, Hong Li, Wei Luo, and Tieyu Wang (2015) 'Impact of soil and water pollution on food safety and health risks in China'. *Environment International* 77: 5–15.

Ma, Jun (16 April 2015) 'Chai Jing'. *Time*. Accessed on 19 July from http://time.com/3823269/chai-jing-2015-time-100/

McDonell, Stephen (24 May 2015) 'China pollution: Beijing's improved air quality a result of good policy, city Officials say'. *ABC.net.au*. Accessed on 19 July from www.abc.net.au/news/2015–05–23/beijing-improved-air-quality-result-of-good-policy-say-officials/6492350

Mufson, Steven (16 March 2015) 'This documentary went viral in China. Then it was censored. It won't be forgotten'. *The Washington Post*. Accessed on 19 July from www.washingtonpost.com/news/energy-environment/wp/2015/03/16/this-documentary-went-viral-in-china-then-it-was-censored-it-wont-be-forgotten/

Ng, Teddy (2 March 2015) 'Chinese celebrity's air pollution video stirs online dust-up'. *South China Morning Post*. Accessed on 19 July from www.scmp.com/news/china/article/1727180/chinese-celebritys-air-pollution-video-stirs-online-dust?page=all

Perry, Elizabeth J. (2007) 'Studying Chinese politics: Farewell to revolution?'. *The China Journal* 57: 1–22.

Perry, Elizabeth J. (2008) 'Chinese conceptions of "rights": From Mencius to Mao – And now'. *Perspectives on Politics* 6(1): 37–50.

Reese, Stephen D. (2015) 'Globalization of mediated spaces: The case of transnational environmentalism in China'. *International Journal of Communication* 9: 2263–2281.

Repnikova, Maria (Spring 2015) 'Media oversight in non-democratic regimes: The perspectives of officials and journalists in China'. *PARGC Paper* 3. Accessed on 19 July from www.asc.upenn.edu/news-events/news/project-advanced-research-global-communication-releases-pargc-paper-3-maria

Rohde, Robert A. and Richard A. Muller (2015) 'Air pollution in China: Mapping of concentrations and sources'. *PLoS ONE* 10(8). doi:10.1371/journal.pone.0135749

Shao, Heng (2 March 2015) 'Only in China: Why a smog documentary sends Chinese stocks soaring to trading limit'. *Forbes*. Accessed on 19 July from www.forbes.com/sites/hengshao/2015/03/02/only-in-china-why-a-smog-documentary-sends-chinese-stocks-soaring-beyond-trading-limit/

South China Morning Post (15 April 2015) 'One dead and 50 arrested after pollution protest in China's inner Mongolia'. *South China Morning Post*. Accessed on 19 July from www.scmp.com/article/1758632/one-reported-dead-after-pollution-protest-chinas-inner-mongolia

State Council (17 March 2015) 'Full text: Report on China's economic, social development plan'. *English.gov.cn*. Accessed on 11 August 2016 from http://english.gov.cn/archive/publications/2015/03/17/content_281475073048566.htm

Tilt, Bryan and Qing Xiao (2010) 'Media coverage of environmental pollution in the People's Republic of China: Responsibility, cover-up and state control'. *Media, Culture and Society* 32(2): 225–245.

Tracy, Megan (2010) 'The mutability of melamine: A transductive account of a scandal'. *Anthropology Today* 26(6): 4–8.

United Nations Environment Programme (2014) *UNEP Year Book 2014: Emerging Issues in Our Global Environment*. Nairobi: United Nations Environment Programme.

U.S. Energy Information Administration (2015) *China: International Energy Data and Analysis*. Washington, DC: U.S. Energy Information Administration. Accessed on 19 July from www.eia.gov/beta/international/analysis_includes/countries_long/China/china.pdf

Wainwright, Oliver (16 December 2014) 'Inside Beijing's airpocalypse – A city made "almost uninhabitable" by pollution'. *The Guardian*. Accessed on 19 July from www.theguardian.com/cities/2014/dec/16/beijing-airpocalypse-city-almost-uninhabitable-pollution-china

Wong, Edward (12 January 2013) 'On scale of 0 to 500, Beijing's air quality tops "crazy bad" at 755'. *The New York Times*. Accessed on 19 July from www.nytimes.com/2013/01/13/science/earth/beijing-air-pollution-off-the-charts.html?_r=1

Wong, Edward (6 March 2015) 'China blocks web access to "Under the Dome" documentary on air pollution'. *The New York Times*. Accessed on 19 July from www.nytimes.com/2015/03/07/world/asia/china-blocks-web-access-to-documentary-on-nations-air-pollution.html

Wong, Edward and Chris Buckley (15 March 2015) 'Chinese premier vows tougher regulation on air pollution'. *The New York Times*. Accessed on 19 July from www.nytimes.com/2015/03/16/world/asia/chinese-premier-li-keqiang-vows-tougher-regulation-on-air-pollution.html

Wu, Fengshi (2013) 'Environmental activism in provincial China'. *Journal of Environmental Policy and Planning* 15(1): 89–108.

Xinhua (29 November 2015) 'China meets pollution reduction targets ahead of schedule: Minister'. *Xinhua*. Accessed on 19 July from http://news.xinhuanet.com/english/2015-11/29/c_134865826.htm

Yang, Guobin (2010) 'Brokering environment and health in China: Issue entrepreneurs of the public sphere'. *Journal of Contemporary China* 19(63): 101–118.

Yang, Guobin and Craig Calhoun (2007) 'Media, civil society, and the rise of a green public sphere in China'. *China Information* 21(2): 211–236.

Zhan, Jiang (2011) 'Environmental journalism in China', in Susan L. Shirk (ed) *Changing Media, Changing China*. Oxford, UK: Oxford University Press, pp. 115–127.

Zhao, Yuezhi (1998) *Media, Market, and Democracy in China: Between the Party Line and the Bottom Line*. Urbana, IL: University of Illinois Press.

5 Mediating risk communication and the shifting locus of responsibility

Japanese adaptation policy in response to cross-border atmospheric pollution

Luli van der Does-Ishikawa
and Glenn D. Hook

Today, East Asian states must respond constantly to the environmental risks and the potential harms posed to their national populations by climate change, externally or internally, or both at the same time. Cross-border atmospheric pollution represents a compelling case of how the external and internal risks and harms produced by climate change are intertwined, not being contained within the boundaries of the sovereign territorial state. The aim of this study of atmospheric pollution, allegedly crossing the territorial boundaries of the Japanese state from China, is to illuminate the processes, implementation and immediacy of Japan's response. Cross-border pollution is of vital concern to state, market and societal actors due to the potential costs and benefits associated with the impact of atmospheric pollutants, especially the risks and potential harms pollutants such as PM2.5 pose to health. Indeed, the harmful effects of the PM2.5 particulate is well established, being known to be carcinogenic, penetrating into the lungs and causing cancer (Sørensen *et al.* 2003; Harrison *et al.* 2004; Vineis and Husgafvel-Pursiainen 2005; Franklin *et al.* 2007; Nakano 2013; Straif *et al.* 2013; Hamra *et al.* 2014; World Health Organization 2014).

Internationally, no foreseeable solution to the risks arising from cross-border pollution has been identified, given the sensitivity of the security and territorial issues between China and Japan as well as the scientific and technological limitations in determining the polluter, notwithstanding judicial and legal hurdles in making any prosecutions. These difficulties go hand in hand with considerations of the economic advantages arising from the trade and financial relations between the two countries. Domestically, a range of risks are posed, headed by the potential harm to the health of Japanese citizens, as evidenced by respiratory diseases, along with the wider impact on the market and society caused by the consequences of cross-border pollution, as illustrated by the reduction in labour productivity and a range of market inefficiencies. The question is: *in order to fulfil its obligations and responsibilities, how does the Japanese state respond in framing the risks and potential harms to the market and society posed by climate change, especially the protection of the people's health in the face of cross-border pollution?*

We develop an original argument that the Japanese state responds to the risks and potential harms posed by cross-border atmospheric pollution by mitigating the potential state–citizen conflict over climate change through a process of policy communication, mediation and implementation using the media, thereby triggering responses from the industrial, societal and other actors whereby the loci of responsibility to protect the citizens cascade down the social structure. This chapter identifies five phases of interrelated responses in this process:

1. The state's communication and mediation of information on the risks and potential harms leading to awareness raising among stakeholders in the market and society;
2. Observation and identification of risks by scientific experts and subnational political authorities;
3. Exchanges between scientific experts and interest groups to explain, legitimise, and to negotiate a consensus on the immediacy of implementing precautionary measures;
4. The manufacturing of protective products and climate-adaptable products in the market;[1] and
5. Mobilising the citizens to shoulder 'self-responsibility' (Hook and Takeda 2007) in responding to the risk to health.

In this process, 'risk discourse' shifts and, in the face of the potential for risks to health being manifest as harms, the risk discourse evolves as if organically as a range of stakeholders and other actors enter a process of negotiation to change the discourse's boundaries. That is, when 'risk discourse' is disseminated by the mass and social media, certain aspects of the discourse are assimilated by market and societal actors, while others become embedded in the given context, creating associations between the original 'risk discourse' and elements of the discourse specific to that context. The communication and mediation of the risk discourse (Hook 2012; Hook *et al.* 2015) in the Japanese media undergoes an assimilation and association process (van der Does-Ishikawa 2013, 2015) of ideological transfer (Valdés 1995) and in this process the hegemony of a 'preferred discourse' (Fairclough 1989) of risk and responsibility becomes embedded. In this way, risks, potential harms and responsibilities are entwined in a process of the agents of the state, market and society negotiating how to respond to cross-border pollution specifically and climate change more generally.

The chapter elaborates how, in this process, the onus of responsibility to respond to the risks and potential harms posed by cross-border atmospheric pollution shifts from the state to subnational political authorities and then to the citizens via industrial and other stakeholders. This process of transition is aided by the dissemination of a discourse of precaution (vigilance), (Dershowitz 2007) and the norm of self-responsibility as well as empowerment to control the impact of atmospheric pollution on health. Our interdisciplinary analysis of the policy documents and media texts issued by the Japanese government reveals in detail how these discourses originate from official documents, but are framed, communicated and mediated by the media. We offer powerful evidence to establish concretely how the state plays the pivotal role in off-loading the responsibility to respond to the risk of climate change to subnational political authorities and other actors, including, crucially, the 'self-responsible' members of society. It is by assimilation and association, we conclude, that the onus of self-responsibility for precautionary measures and risk alleviation is assigned to these subnational authorities and then to the citizens. The process is one of mediating risk communication in support of shifting responsibility from the state to other actors – the off-loading of state risk as part of the wider process of the shrinking of the Japanese state (Hook and Takeda 2007). The data analysis described below explores this process in the risk-responsibility matrix in the context of 'cross-border air pollution' as reported by the Japanese media, using a combination of in-depth quantitative assessment and qualitative analysis.

To be more specific, the chapter offers a theoretically informed empirical study of the Japanese media discourse on cross-border air pollution between 1998 and 2016. It seeks to uncover the time-varying patterns of pollution with a specific focus on the reporting of PM2.5, a particulate identified by the Ministry of Environment as one of the potential harms to the health of the citizens posed by cross-border atmospheric pollution (Ministry of Environment, Japan 2014a). It does so by examining the reporting on this pollutant in the national and regional broadsheets published in Japan, both paper and online, restricted to outputs in the Japanese language.

We proceed by dividing our discussion into four sections. This first section describes the structure of the study. Section two briefly introduces our theoretical approach in order to explain the processes by which policy is disseminated with the aid of media discourse. It then delineates the concept of the Japanese government's 'adaptation policy' and spells out the hypotheses and analytical methods deployed in the empirical analyses. Section three then moves on to analyse the results of the empirical study, involving three distinct stages. This is the longest section and is composed of a number of interrelated parts. First, the data are elaborated and their sufficiency confirmed. Second, the frequency of media coverage of cross-border air pollution in Japan in the time frame of August 1998 to March 2016 is analysed in order to elucidate the temporal dynamics of reporting by identifying the peaks in the number of news articles that report on PM2.5. This process yields nine periods – or 'waves' – of distinct news foci. Third, the news contents of each of these nine periods are scrutinised to reveal nine distinct discourse patterns and shifting news foci from one period to the next. In each of these nine periods, the patterns of text-internal relationships between the concepts of 'risk', 'responsibility' and cross-border air pollution are codified and statistically explored based on what we believe is an original analytical framework and set of hypotheses. The findings are then linguistically evaluated against the historical background. In particular, the study focuses on *how* things happen, enabling us to demonstrate how the concepts of 'risk' and 'responsibility' interact with the concept groups of 'actors', 'actions', 'sources (of pollution)', and then to move on to show how such interrelationships may have changed over time in mediating risk communication and embedding specific media discourses of cross-border pollution in the market and society.

In this process, the chapter sheds light on how the concept of 'responsibility' is assigned to a particular actor at a given time, and how the locus of assigned responsibility shifts from one actor to another over time in a given context of domestic and international, environmental and political, contexts. Section four then takes up the empirical results of section three within the theoretical framework of risk and responsibility (Hook 2012; Hook *et al.* 2015) to identify the roles played by the media in the process of risk communication, mediation and the allocation of responsibility among the state, market and society under a set of specific parameters of actions, actors and the sources of cross-border air pollution. The conclusion offers a tentative process model of risk communication and shifting responsibility as the basis for further research.

The 'adaptation policy' and hypotheses

Japan's 'adaptation policy' originates in the Intergovernmental Panel on Climate Change's (IPCC) Fourth Assessment Report issued in 2007 (Pachauri and Reisinger 2007), which emphasises the importance of both mitigation and adaptation strategies in order to deal with the challenges of climate change. Mitigation strategies are crucial in slowing, stopping and reversing the accumulation of greenhouse gases (GHGs). Mitigation strategies may include the direct reduction of GHG emissions, while providing economic and technological resources necessary for other stakeholders (e.g. poorer developing nations) to reduce emissions, and preserving 'carbon sinks' (e.g. rainforests) able to absorb GHGs that otherwise would accumulate in the atmosphere. In contrast, adaptation strategies focus on avoiding the harmful consequences of global climate change, and typically involve changes to human behaviour or sociopolitical systems that impact on the environment. These are put forward as mutually complementary twin strategies to tackle climate change as in the key message of the IPCC's Fifth Assessment Report on adaptation-mitigation policy implications (Edenhofer *et al.* 2014). The actual definition of their contents and ways of implementation are left to the member-states, however. In this sense, room remains for states to *reinterpret* these international efforts.

The adaptation policy

Prime Minister Abe Shinzō's speech at the Plenary Session of the United Nations Climate Summit in September 2014 alerted the world to the Japanese government's 'adaptation initiative' adopted in response to climate change in which 'mitigation strategies' were not mentioned (Abe 2014; IISD 2014).

It offers a Japanese perspective on the initiative advocated in the 2007 IPCC report on Climate Change Adaptation and Mitigation for Sustainable Risk

Management (Denton *et al.* 2014). IPCC members, including Japan and the countries of the European Union (EU), have adopted the initiative in domestic policies to varying degrees. As an example, fifteen EU member-states, including the United Kingdom, adopted a national adaptation strategy in January 2013, and by April of the same year EU-wide guidelines for an adaptation strategy were issued as a way to increase the level of coordination and standardisation, emphasising the importance of disseminating accurate and up-to-date climate information behind the policies as well as stressing the need for a concerted effort in response by all stakeholders. In other words, the response to climate change at this temporal juncture was to adopt an approach based on *shared responsibilities* among the member-states, internationally, as well as *shared responsibilities* within the member-states, domestically, embracing a range of actors, such as subnational political authorities, industrial actors in the market as well as citizens in society.

In response to the IPCC reports, Japan introduced an adaptation policy to combat climate change. A notable example is the launch of the initiative called 'Proactive Foreign Policy Strategy against Global Warming' (*Seme no chikyū ondanka gaikō-senryaku*), also known as 'ACE: Actions for Cool Earth' by interministerial coordination (Ministry of Foreign Affairs, Ministry of Economy, Trade and Industry and Ministry of the Environment of Japan 2013). In fact, the ACE initiative was originally proposed during the first Abe administration in 2007 and revamped in 2013 during his second term in office (2012–). The initiative's long-term goal is to reduce the total volume of GHG emissions in the developed countries by 80 per cent and halve the world's total GHG emissions by 2050, while a key short-term objective is to lead the domestic and international negotiations on efforts to combat climate change towards designing an updated United Nations (UN) Framework on climate change negotiated during the UN COP21 held in Paris in December 2015. This led to an agreement among 195 countries to keep the rise in global temperature below 2 degrees Celsius (United Nations Framework Convention on Climate Change 2015). The Japanese initiative consists of three pillars: (technological) innovation, application and intergovernmental partnerships, whereby the 'innovation and application' of technologies focuses on revitalising domestic industries through innovative environmental businesses and intergovernmental partnerships comprise economic and technical assistance to developing countries through Official Development Assistance (ODA), other Official Flows (OOF), as well the use of funds from private-sector market actors. In this way, the three pillars of the Abe administration's adaptation policy to tackle climate change have a strong connotation of economic benefits centred on the role of the market.

As regards the communication of information on climate change and the 'adaptation policy', the Ministry of Environment has held symposia and meetings addressing the public since 2011 in order to 'disseminate the Ministry's research findings widely to all quarters of Japanese citizens while promoting local municipalities' research on the effects of climate change and adaptation and contributing to their adaptation policies'[2] (Ministry of Environment, Japan

2012a). Implementation of the 'adaptation policy' was included in the ministry's priority measures for 2013 (*Heisei 25-nendo Kankyōshō Jūten-sesaku*) (Ministry of Environment, Japan 2012b), and a comprehensive report, 'Efforts of the Ministry of Environment, Japan, towards Adaptation to Climate Change' (*Kikō-hendō no tekiō ni muketa Kankyō-shō no torikumi ni tsuite*), was presented at the second symposium organised by the ministry (Ministry of Environment, Japan 2013b) on 26 November 2013. Concurrently, a series of locally based symposia was held, entitled 'Towards IPCC 38th. Symposium to Consider the Impact and Solutions of Climate Change Close to Home' (*Kikō-hendō no mijikana eikyō to tekiō-saku wo kangaeru shimpojiumu IPCC dai-38kai sōkai ni mukete*) across the nation starting from Kyoto Ministry of Environment, Japan 2013c). After the IPCC 38 in May 2014, a new series of symposia entitled, for example, 'Climate Change Sciences and our Future: Dialogue between IPCC and the Citizens of Fukushima Prefecture' (*Kikōhendō no kagaku to watashitachi nomirai – IPCC to Fukushima-kenmin no taiwa*), was held across Japan (Ministry of Environment, Japan 2014c). The ministry's dissemination activities reveal in identifiable form the shift taking place before and after the IPCC 38: namely, the focal point has moved from communicating the responsibility of the state and the efforts made by the Abe administration in response to climate change to the benefits of science and technology in contributing to the protection of the ordinary citizen's health and everyday life in the face of the risks and potential harms of climate change. The *shared responsibilities*, both international and domestic, as emphasised in the EU Climate-ADAPT strategy (Hill and Smith 2000: 233), are now far less clear-cut in the ministry's official documents.

Notably, the analysis described in this chapter offers detailed evidence of how the generic first- or third-person pronouns are employed – or implied by the context – ubiquitously, indicating the generic 'person' as the locus of responsibility to respond to climate change. But this does not mean that the locus of responsibility remains unclear domestically in Japan: rather, we argue that, over time, the locus of responsibility to implement the adaptation strategy in response to climate change has been shifting from the state to subnational political authorities, as well as other actors in the market and the society, through a gradual, covert persuasion communicated through the media. In this process, the state appears to be off-loading risks to subnational political authorities, the market and societal groups via emphasis on precautionary measures and self-responsibility, while at the same time taking state-level responsibility at the level of international society (Hook 2012). Based on these premises, the hypotheses put forward below aim to test the validity of these views.

Hypotheses

The theory of risk communication that is pertinent to this study is the social amplification of risk framework (SARF), proposed by Kasperson *et al.* (1988). It provides a conceptual model to explain the motivational factors of mediated dissemination of risk discourse. Certain aspects of hazardous events and their portrayal in the media and other sources interact with *psychological, social, institutional and cultural processes* in ways that might attenuate or amplify perceptions of risk (Kasperson *et al.* 1988: 177). Through this, risk communication shapes behaviour, as in the case of climate change debate (Renn 2011). Taking the SARF further, we propose a tentative empirical and theoretical model to explore the processes of discourse dissemination in media whereby the loci of risk and responsibility *shift*.

Risk is a tool of governance, where precautionary measures have become integral to the pattern of behaviour adopted by the citizens in the process of being governed (Hook 2012: 324). This is part of a wider process of taking '(self-) responsibility' in responding to risks through recalibrating citizens' exposure to external and internal risks and potential harms (Hook and Takeda 2007: 93). In order for the governed citizens to adopt such a pattern of behaviour, however, they must first be exposed to and become familiarised with the notion through assimilation and association of the discourse patterns as communicated and mediated by the media (van der Does-Ishikawa 2013). The first step, then, is to disseminate information on risks and embed them in the context of climate change (Hook *et al.* 2015: 42–43). Next comes legitimisation (Lester and Hutchins 2014: 859), consensus-building (McCombs 1997) and finally mobilisation (Lester and Cottle 2009), whereby media coverage plays a pivotal role in the 'willingness and capacity' of policymakers and citizens to 'respond and act' (Lester and Cottle 2009: 929). Thus, responsibility is off-loaded from the state to society as well as to the market, as illustrated below in the role of the citizens in shouldering the responsibility to respond to cross-border atmospheric pollution.

The risks and potential harms to health posed by PM2.5 are communicated through the media to 'raise the awareness' of the citizens. Meanwhile, scientific experts discuss the health and other risks posed, highlighting the risk's relevance at the societal level and the potential harms to the everyday security of the individual citizens (Nagashima 2015). Increased awareness leads to a process of potential conflict and negotiation whereby the citizens call on the state to take actions to counter/mitigate the risks to health posed by climate change. The adaptation policy endorsed by the government is implemented in this context. A wide range of stakeholders and other actors are involved in the processes of policy implementation, including the state, subnational political authorities, relevant market actors such as industries, as well as specialist groups (scientific and medical experts, etc.), interest groups and ordinary citizens. Meanwhile, the state authorities and experts advise local communities to take care-based measures or provide prudence-based precautionary guidelines. These dual activities result in a shift in the locus of responsibility for action, from the state to subnational authorities, the market and society, with the norm of 'self-responsibility' being inculcated at the individual level.

We would expect evidence of the above to be manifest in the Japanese media discourse of 'cross-border atmospheric pollution', which is characterised by its frequent reporting on 'risks', and presents a range of shifting emphasis on a combination of 'source/cause', 'risk', 'responsibility', 'action (including policies)' and 'actors' over time. This scenario is reformulated as the following set of hypotheses:

1   Frequency of media reporting on the topic of PM2.5 and the focus on a particular aspect of the topic change over time coinciding with a relevant sociopolitical and environmental event.
2   News discourse on different aspects of cross-border air pollution exhibits distinct discourse patterns characterised by different lexical groups.
3   Different media discourse patterns observed at different time periods shed light on how different actors, actions and sources of hazard interact to instantiate that discourse at a given time.
4   Shifts in the pattern of discourse across periods display changes in the association between certain actors, actions and sources of hazards (PM2.5).
5   Shifting discourse patterns on risk and responsibility over time display a phenomenon of top-down 'off-loading' of responsibility from the state to other actors coupled with the channeling of momentum into individual actions based on 'self-responsibility'.[3]

These five hypotheses are tested below using a combination of statistical and discourse analytical methods. The results are reported and discussed both quantitatively and qualitatively.

Study on risk and responsibility: data, methods, results and discussion

Data

Data for this study consist of the texts of news articles and their attributive information collected over the period from August 1998 to March 2016, which forms our master database. All news articles were collected using a restrictive and specific term from the four major news search engines, namely, Nikkei Telecom 21, Yomidas, Kikuzo and Mainichi, each representing the four major Japanese language broadsheet news companies, Nikkei, Yomiuri, Asahi and Mainichi, respectively.

Various work data subsets were extracted through stratification to serve different analytical purposes, such as demographic and text content analyses. The master database records article titles and texts along with attributive information such as dates of publication, publishers, textual contents and other defined categories on Microsoft™ Excel® worksheets so that the data will be readily available for subsequent use with specific searchable database languages and/or software.

All the data used for this study were retrieved by a keyword search with the terms 'PM2.5', 'PM2.5-Risk' and 'PM2.5-Responsibility'. The search returned a total of 917 articles. Of these, 126 were published in the Japanese language as an adaptation or translation from another source language (i.e. translations from non-Japanese newspapers or press agencies articles such as 人民網 [*People's*

Daily] or Dow Jones), and therefore they were excluded to ensure the homogeneity of the data origin. As a result, 791 articles comprising 1,172 paragraphs were subjected to analyses. Within this set, articles conveying the word 'risk' were 280 (35.40 per cent) and 'responsibility' 166 (20.99 per cent), whereas those conveying both words only amounted to 21 (2.65 per cent). The apparent paucity of the usage of 'risk' and 'responsibility' is because Japanese professional journalists may not frequently repeat these topical words, but are more likely to opt for their paraphrases for the sake of literary variety and depth.[4] Therefore, it is crucial to assess these key concepts through semantically related word groups. Hence, as shown below, we propose an original methodological approach to incorporate coding rules into an analysis of co-occurrence network,[5] while contextualising the interpretation of the results using the techniques of critical discourse analysis.

The search returned articles from eight different paper categories: national news, local news, news service agencies, industrial news, sports news, internet news, news of public origin (e.g. interest groups) and other. Given the variety of news sources, the homogeneity of the textual characteristics of the data set was statistically examined as shown in Table 5.1. The number of news articles per newspaper category is also shown in Table 5.1.

Analytical methods

Our analytical approach combines a mix of quantitative and qualitative methods in order to achieve a transdisciplinary dimension to the research through the well-known benefits of triangulation drawing on social sciences, applied linguistics and Bayesian as well as frequentist statistics. The process of transient assignment and reassignment of risk and responsibility to specific agents in the discourse on environmental adaptation is thoroughly assessed and analysed, employing distinctively innovative interdisciplinary analytics as described in the following lines.

Descriptive statistics

First, a histogram of news frequency on PM2.5 was compared to the timeline of historical events to give an overview of synchronicity between the politico-environmental events and the media response between 1998 and 2016. The graphs reveal the peaks and troughs of media interest in different aspects of PM2.5, shifting from one aspect to another over the nine historical periods identified below.

Correspondence analysis (CA)

To summarise the shifts in the texts' characteristics over the course of the nine periods, correspondence analysis (CA) was used as an exploratory technique to visualise the relationship between individual words and the time periods. The resulting biplot indicates a clear pattern of shifts. CA is a statistical method derived from

Table 5.1 Homogeneity of the text by types of news providers*

Category	No. of articles	Mean	SD	Min.	Max.	Kurtosis	Skew-ness	Chars >4,000	% of long txt
National news	325	929.22	920.03	121	7,792	15.10	3.13	4	1.23
Local news	297	747.71	656.52	114	5,359	16.43	3.31	2	0.67
News services	55	616.16	377.55	134	2,450	9.01	2.23	0	0.00
Industrial news	46	1,307.61	1,785.34	164	**10,875**	18.88	4.00	3	**6.52**
Sports news	6	812.67	673.24	281	1,980	0.70	1.37	0	0.00
Internet news	5	1,544.00	1,384.94	272	3,836	2.43	1.47	0	0.00
Other	18	1,059.39	1,502.63	191	5,922	7.13	2.75	2	11.11
Public origin	39	1,789.49	2,515.54	188	**13,650**	13.10	3.30	5	**12.82**
All articles	791	909.69	1,063.99	114	13,650	42.27	5.17	16	2.02

Source: van der Does-Ishikawa (2015)

* The maximum number of characters is very high in the news from industrial and public origins. Also, the percentage of long texts are higher in these news types. Generally, business publications and news issued by public bodies tend to be lengthy and more detailed in this data set. However, the purpose of this study is not to compare the news contents among the same type of publications, but to explore the general trend in the media discourse of cross-border air pollution in all original news articles published in Japanese language during a specific time frame. Therefore, this data set is deemed appropriate for the purpose of this study and for the hypotheses tested.

Pearson's principal component analysis. It is a dimension reduction of categorical variables: CA applies χ^2 statistics on a contingency table, or cross-tabulation, and then maps the results over a biplot to visualise the interrelationship between data sets over orthogonal coordinates. The technique provides a summary of the data for exploratory purposes, without probabilistic evaluation in regard to the relevance of a particular aspect of the data. In other words, CA in this study offers a visual aid as a starting point to explore our data set (Yelland 2010).

Multidimensional scaling (MDS)

Another exploratory method is multidimensional scaling (MDS). The multivariate analysis reveals the underlying semantic structure of a data set intuitively through visualisation. Frequently occurring words are classified into clusters according to their strength of association or similarity, and their interrelationships within the texts at multiple layers are reinterpreted for a simplified representation in the form of a two- or three-dimensional graph. In our analysis, MDS reduces the multiple dimensions into two. Although this leads to some degree of approximation in the resulting plot, it proves useful at the same time to gain an overview of the texts' internal semantic relationship that gives the reader an overall impression of the news articles. Thus, we identify how the media represents the news of PM2.5 using several semantic groups. The principles are briefly described in the following lines.

Assuming that, to map n words (1 . . ., n), for a word $i \neq j$, their number of co-occurrences is c_{ij}, whereby $c_{ij} = c_{ji}$ and the total number of co-occurrences of a word i is c_i:

$$c_i = S_{j \neq i}\, c_{ij}$$

In this study, we use the widely popular cosine as the similarity measure. The cosine between items i and j can be expressed as (van Eck *et al.* 2010):

$$\text{COS}_{ij} = \frac{\sum_{k \neq i,j} c_{ik} c_{jk}}{\sqrt{\sum_{k \neq i,j} c_{ik}^2 \sum_{k \neq i,j} c_{jk}^2}}$$

To note, the hierarchical cluster annexation plot was used as the background data to determine the minimum number of clusters.

Co-occurrence network

The co-occurrence network analysis was employed to identify textual characteristics of media discourse by examining the word frequency over the nine historical periods, whereby the temporal factor (i.e. the nine periods) was introduced as an external variable. Co-occurrence analysis examines word frequencies and their co-occurrences within texts, calculates how closely the words are related and visually represents the distance and proximity of interrelations by drawing 'edges'. The strength of relationships between and among these words is expressed by their similarity, corresponding to geometric measures of vector closeness in a network graph. In this study, we use the KH Coder default Jaccard coefficient as the similarity measure. While there are other measures or research instances where normalisation is necessary, the Jaccard coefficient remains one of the most popular set-theoretical measures of similarity, and has the property to focus on strong links.

The Jaccard coefficient can be expressed as follows:

$$J(A,B) = \frac{|A \cap B|}{|A \cup B|}$$

or, in other words, as a similarity index S_J of items k and j

$$S_{Jkj} = \frac{w_{kj}}{w_{kk} + w_{jj} + w_{kj}}$$

where w_{kk} is the total number of occurrences of k and w_{kj} is the number of co-occurrences of k and j.

Coding rules

As discussed earlier, Japanese professional media discourse deploys a complex web of paraphrases when referring to a key concept, rather than simply repeating the same word. In order to take this into account, we have incorporated coding rules into the analysis of co-occurrence networks. The original coding scheme was developed using paraphrases identified and grouped within the list of most frequently occurring words returned by the quantitative analysis. Weblio (2015) served as one of the online reference dictionaries-thesauri to set and check against these groups, referred to as 'codes', of words in this study. The 'coding' scheme consists of three main topical categories (cross-border air pollution, risk and responsibility), three fact-oriented categories (actor, action and source) and their respective subcategories. The coding scheme was verified by an intercoder reliability test (De Swert 2012). Articles that contain words consisting of these codes are subjected to analyses. The number of articles containing the members of the code for 'risk' was 479 (60.56 per cent), 257 (32.49 per cent) hits for 'responsibility', and 168 (21.24 per cent) hits that contain members of both codes.

The relationships between the codes and the time periods, introduced as an external variable, have been formulated in a contingency table between the external variable and the categories of coding rules. Potential relationships between the keywords in the internal structure of the media texts are graphically visualised using co-occurrence networks for analyses.

Heatmap

Various parameters are expressed in a contingency table and their interrelationship between the cross-tabulated variables (actors, actions and sources), on the one hand, and the nine periods, on the other. The strength of the relationships is visualised by the intensity of shades; that is, the more intense shade the stronger the relationship.

Critical discourse analysis

Qualitative analysis of the news articles is performed concurrently with the quantitative analyses using the cognitive, process-oriented branch (Koller 2014) of critical discourse analysis (Wodak and Meyer 2009).

Computerised analytical tools are employed to aid the statistical analyses.

Tools

Text analytics is performed using the IBM SPSS version 22 and the Japanese-compatible text mining tool KH Coder, a package used in more than 1,200 academic peer-reviewed papers to date (Higuchi 2004: 101–115). The package deploys a conventional part-of-speech (POS) scheme, such as noun, verb, at lexical and/or morphological levels using a set of rich back-end tools and programming languages, including R, Stanford POS Tagger, Snowball stemmer, Python, MySQL, ChaSen (for Japanese) and MeCab (for Japanese). The last two parse the Japanese text by morpheme analysis, using Markov chain Monte-Carlo (MCMC) for the former and hidden Markov model for the latter, each with their own specific compatible dictionary format, IPADic and UniDic, respectively. While the dictionaries available for the former are suitable for modern Japanese, those for the latter offer means for the analysis of Meiji to mid-twentieth century Japanese, thus we have used the IPADic principally for this study. The current stable version 2.0 handles Japanese, English, French, German, Italian, Portuguese and Spanish through different combinations of back-ends, and the next version 3.0, still under development at this date, is expected to be able to handle both Korean and Chinese, as well as Serbian, among others.

Shifting topics and changes in the number of publications

The first hypothesis to test is:

(a) Frequency of media reporting on the topic of PM2.5 and the focus on a particular aspect of the topic change over time coinciding with a relevant sociopolitical and environmental event.

The results show that news articles belong to two distinct historical periods: August 1998 to January 2013, on the one hand, and February 2013 to March 2016, on the other. The number of articles on PM2.5 cross-border atmospheric pollution, which remained at less than five per month until January 2013, shot up in February 2013. That is, the maximum number of publications is markedly different before and after January 2013.[6]

A closer examination of Figure 5.1 reveals eight peaks in the number of publications, each of which is followed by a steady slope of decline. As news material tends to peak when news is 'new', and the level of coverage tends to decrease as the novelty diminishes and the interest of the readership wanes, we

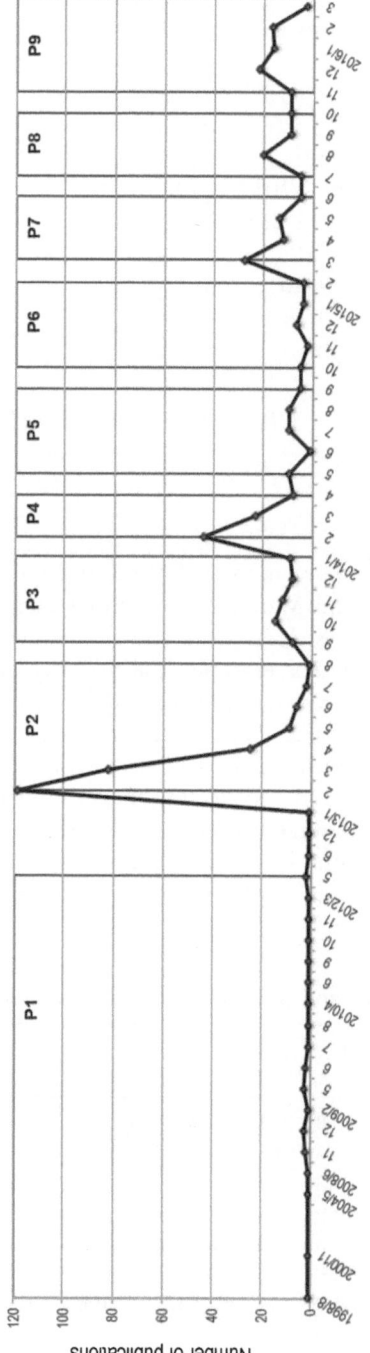

Figure 5.1 Number of publications on PM2.5, risk and responsibility

Source: van der Does-Ishikawa (2016)

consider these peaks as the beginning of the cycle of a new 'newsworthy' topic, and divide the overall period into nine subperiods:

Period one: 1998 August – 2013 January,
Period two: 2013 February – 2013 August,
Period three: 2013 September – 2014 January,
Period four: 2014 February – 2014 April,
Period five: 2014 May – 2014 September,
Period six: 2014 October – 2015 February,
Period seven: 2015 March – 2015 June,
Period eight: 2015 July – 2015 October, and,
Period nine: 2015 November – 2016 March.

Once again, the homogeneity of textual characteristics is examined, yielding the results shown in Table 5.2.

Historical analysis

The historical record of events relating to PM2.5 and cross-border air pollution between August 1998 and March 2016 demonstrates a striking finding. That is, a set of events coincides closely with the boundaries between *every* one of the nine periods identified. For example, the majority of articles published during period one report on air pollution caused by factories and the emissions from automobiles, expressing concern about the increase of atmospheric pollution worldwide and particularly in China, occasionally referring to the scientific studies on PM2.5, and recollecting Japan's own past struggles in response to pollution. The coverage represents a gradual introduction of the scientific findings on the harmful effects of PM2.5, very occasionally reporting information in relation to China's industrialisation and the consequential risks

Table 5.2 Distribution of texts and their length over the nine periods

Period	No. of articles	Mean	SD	Min.	Max.	Kurtosis	Skewness	Chars > 4,000	% of long txt
1	41	1,281.44	1,230.78	291	7,617	17.39	3.59	1	2.44
2	280	1,022.68	805.73	165	5,212	7.44	2.40	4	1.43
3	98	1,130.56	1,692.83	134	1,3650	31.35	4.90	6	6.12
4	105	843.69	992.69	121	7,792	25.99	4.51	2	1.90
5	48	755.42	888.13	164	5,359	15.21	3.47	1	2.08
6	50	653.60	1,548.89	153	10,875	40.63	6.16	1	2.00
7	59	563.00	446.98	124	2,042	2.63	1.76	0	0.00
8	42	715.31	1,040.70	114	6,710	27.97	4.93	1	2.38
9	68	721.94	686.51	140	3,694	7.04	2.48	0	0.00

Source: Luli van der Does-Ishikawa (2016)

of air pollution, along with the possible impact of pollution on health. Such general coverage, however, shifts suddenly in February 2013. That is, between 30 January and 2 February a number of regional newspapers start to report on the prolonged atmospheric concentration of PM2.5 exceeding 250 µg/m³ in China and the risks of cross-border air pollution in parts of Kyushu prefecture, but by 4 February nearly all articles mentioning PM2.5 now point to the risks of cross-border atmospheric pollution from China in a 'China threat' narrative framework (Hook *et al.* 2015: 43) and the urgent need for the state to respond by adopting a policy to combat the potential harms.

At the time, experts on air pollution and environmental policymakers were obviously aware of the issue of PM2.5, as Japan's Ministry of Environment held a symposium called 'Climate Change Sciences and our Future: Discussions with the IPCC Bureau and Authors' (*Kikōhendō no kagaku to watashitachi no mirai – IPCC Gichōdan, shippitsusha wo mukaete*) in Tokyo on 29 January 2013. Indeed, the date is precisely when the sudden explosion of interest in cross-border air pollution began in period two. During the symposium the Abe government's ACE initiative was trumpeted as Japan's action plan in response to the IPPC's 'adaptation and mitigation' policy on climate change (Ministries of Foreign Affairs, Economy, Trade and Industry and Environment, Japan 2013). A report of the committee of experts on PM2.5 was issued on 27 February recommending actions in response to the potential harms of atmospheric pollutants, following on from which the Ministerial Ordinance on the 'Partial Revision of the Ordinance for Enforcement of the Air Pollution Control Act' (*Taikiosen bōshi sekō kisoku no ichibu wo kaiseisuru shōrei*) was issued on 3 March. On 29 March the legislation for the 'Act for Partial Revision of the Air Pollution Control Act' (*Taikiosen bōshi-hō no ichibu wo kaiseisuru hōritsu-an*) was approved by the Abe cabinet. These developments coincided with the high news output on PM2.5 throughout February and March in period two, before coverage tails off towards and throughout the summer.

The next sudden surge in reporting, though smaller than in period two, occurs in September 2013 at the beginning of period three. This time the news topics revolve around Beijing's announcement of the government's action plan to combat air pollution (State Council of the People's Republic of China's 'Action Plan on Prevention and Control of Air Pollution'), but highlighted China's perceived 'disinterest' in investing in a low-emission policy. For the Japanese media, the Chinese response was taken as evidence that the risks from cross-border air pollution can be expected to continue unabated. The 2013 report of the World Health Organization on the potential harms to health from the risks of PM2.5 (World Health Organization 2013) is cited frequently, and a concrete policy to protect the citizens is called for in response. The Ministry of Environment, Ministry of Labour, Health and Welfare and Ministry of Economy, Trade and Industry, among other ministries, soon organised a series of expert committees to draw up policies for implementation. For instance, the Ministry of Environment made an announcement on Japan's commitment to implementing an adaptation policy targeting climate change, while referring to the IPCC Fifth

Report (Working Group I) and uploaded a full translation onto the ministry's website (Ministry of Environment, Japan 2013a).

A steady level of news on cross-border air pollution continues throughout period three until January 2014, during which COP19-CMP9 was held on 11 November in Warsaw. It is here that the Abe government announced a revised short-term target to reduce carbon dioxide (CO_2) emissions. Within a week, on 15 November, the Japanese cabinet had revised the ACE adaptation policy, and the Ministry of Environment announced the update of 'climate adaptation' actions on its website on 26 November. The ministry then organised a series of symposia, targeting regional stakeholders and other actors involved in policy implementation in response to cross-border air pollution, such as representatives of subnational political authorities. The symposia started on 29 November and were held in key regions of the country until the end of January 2014. A policy package on PM2.5 was issued on 25 December. News articles in period three often contrast Japan's policies with those adopted in response to air pollution in China, the United States and the European Union.

Period four begins in February 2014 when the next upsurge in coverage of cross-border air pollution occurs. The media equivocally report on what is often referred to as the 'annual occurrence' of exceedingly high concentrations of PM2.5 in China. Regional newspapers begin to report frequently on the observed level of cross-border air pollution and the environmental challenges faced by China. NHK's special issue, "How should we face the cross border air pollution?" (*Ekkyō taiki osen to dō mukiau ka*), was aired nationwide on 22 February 2014 ahead of the Ministry of Environment's press release on the IPCC Fifth Report (Working Group II) and its translation with a summary for policymakers. Linking to the IPCC findings, the Fourth Strategic (Fundamental) Energy Plan was issued on 11 April 2014 followed by a press release on the IPCC Fifth Report (Working Group III) on 13 April (Ministry of Environment, Japan 2014b). Throughout period four regional newspapers reported the observed levels of cross-border air pollution, scientific approaches, health risks, as well as the environmental policies of China and Japan. The frequency of reports tails off over the course of April.

The Sixteenth Tripartite Environment Ministers Meeting among South Korea, Japan and China (TEMM16) was held on 28–29 April (Ministry of Environment, Korea 2014), and news of the event appears to trigger the next surge of news in period five starting in May 2014. National and regional newspapers reported observed high levels of atmospheric PM2.5 concentration. Regional newspapers reported that the observed levels were exceeding, or expected to exceed, the standards set by the Ministry of Environment for exposure to fine particulates in the atmosphere. News topics revolved around national and international collaboration through public and private partnerships to 'adapt' to climate change. The volume of news on PM2.5 decreases over the summer.

At the end of September to the beginning of October 2014 the frequency of news on PM2.5 spikes again, marking the beginning of period six. This time, the news concentrates on international events such as Prime Minister Abe's speech at the United Nations Climate Summit in New York (Ministry of Foreign Affairs 2014). The risks and potential harms of air pollution emerged again, as illustrated by the concern for the health of the runners participating in the annual International Marathon event in Beijing in October and the 2014 meeting of the APEC forum held in China on 10–12 November 2014. The issue of climate change was often linked to concerns about energy supply during this period.

Periods seven to nine manifest smaller spikes, but still show distinct peaks and troughs, where the peaks of March 2015 (period seven), August 2015 (period eight) and December 2015 (period nine) are associated with news on environmental policies. Period seven reports Japanese and Chinese domestic policies; period eight discusses China's legal framework and Japan–China research collaboration; and period nine reports on the COP21 and China's serious atmospheric pollution. Here again, a gradual shift in the attention devoted to different aspects of PM2.5 can be observed.

The shifts in discourse patterns across the nine periods are clearly evidenced from the above. We therefore conclude that hypothesis (a) is supported by the results. As these periods differ from each other, as hypothesised, presenting distinct discourse patterns, this brings us to the next hypothesis:

(b) News discourse on different aspects of cross-border air pollution exhibits distinct discourse patterns characterised by different lexical groups.

To test hypothesis (b) the text-internal structure of lexical contents of news articles across the nine periods is explored. The results are shown in Figure 5.2 in which different periods display distinct lexical sets with period six as an obvious outlier.

A detailed observation reveals that the concept of risk (*risuku*) is most closely associated with periods two and three and such concepts as health (*kenkō*) and environmental standards (*kijun*), while the concept of responsibility (*sekinin*) is associated with periods three and possibly periods one, five, seven and nine. Concepts associated with responsibility appear to be regional (*chiiki*), action (*okonau*), management (*kanri*) and protection (*hogo*), among other things. This finding points to period-specific characteristics of news contents whereby concepts of risk and responsibility may be associated with a particular media content of a specific time period. This analysis is still too crude and calls for further study, as discussed below.

We now investigate the characteristics of the news articles in each of the nine periods in order to probe if news contents changed over time, displaying distinct time-varying patterns. To explore the semantic patterns of media output in each period, co-occurrence analysis is employed to test hypothesis (b) further. The results are presented in the next section.

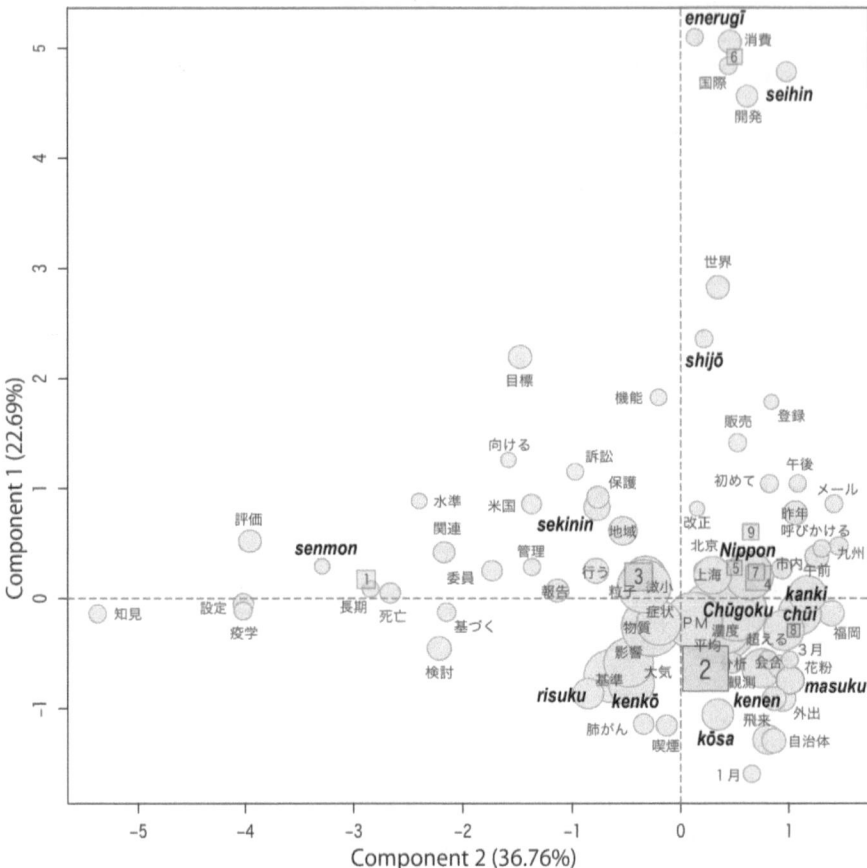

Figure 5.2 Correspondence analysis of the frequently occurring words over the nine periods

Source: van der Does-Ishikawa (2016)

Shifting semantic contents in the media over the nine periods

To start with, word frequency was analysed within the data set obtained using the keywords, 'PM2.5', 'risk', and 'responsibility'. These were then subject to co-occurrence network analysis. The results are visualised in Figure 5.3, which shows the shifting semantic contents of periods one to nine.

As regards the convention of the graphs, items within the same subgraph, or community, are coded in the same shade of grey, visualising the relationship between words that co-occur. The periods are shown as indicated by their number in squares. Words predominant and unique in a given period are shown in white circles, while those shared between periods are displayed in dark- and light-gray circles. The thickness and the tone of the edges (i.e. the lines connecting the

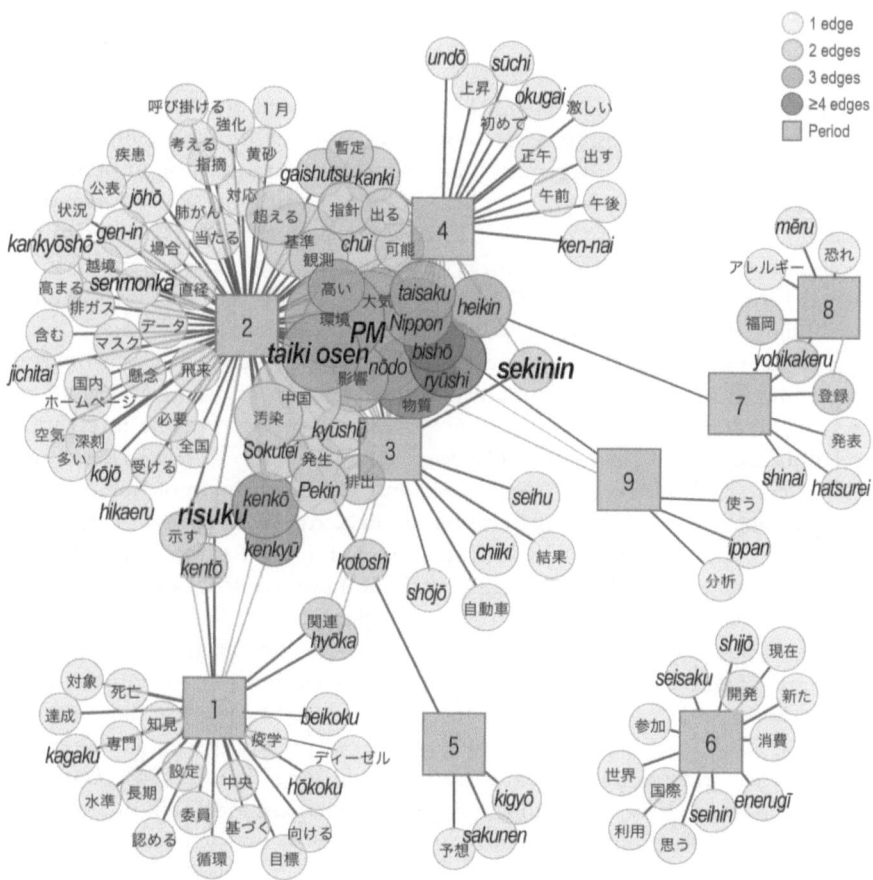

Figure 5.3 Co-occurrence of frequently occurring words in the risk and responsibility data – all nine periods

Source: van der Does-Ishikawa (2016)

different words and nodes) depict the strength of the connections whereby the stronger the connection, the thicker and darker the *edge* (line). Note that, contrary to CA and MDS, the distances separating the words are not relevant here, the importance being the centrality and the strength of the links between and among words.

The co-occurrence network presents nine distinct communities for each period, with eight out of the nine communities interlinked through at least one node. Compatible with the result of correspondence analysis discussed earlier (Figure 5.2), period six is entirely independent from any other communities. We therefore conclude that the news contents of this period are quite distinct from the other eight periods.

Period one shares nodes with periods two and three in light and dark shades of grey. This period is characterised by words that report (報告 *hōkoku*) the past experience and knowledge of specialised scientific studies on fossil-fuel pollution (ディーゼル *dīzeru*, 専門 *senmon*, 疫学 *eikigaku*, 知見 *chiken*, 科学 *kagaku*, etc.), suggesting that before January 2013 the link between the concept of PM2.5 and cross-border atmospheric pollution was not yet firmly established. This accounts for PM2.5 being reported in a generally tentative tone in period one. For example, pollution is evaluated (評価 *hyōka*) based on (基づく *motozuku*) the US (米国 *beikoku*) studies (研究 *kenkyū*) that report death (死亡 *shibō*) from pollution, and relevant (関連 *kanren*) parameters (設定 *settei*) of risks (リスク *risuku*) are reviewed (検討 *kentō*). Period one relates to period two through the small light-grey community of words that signify evaluation of risks.

Periods two, three and four are intricately linked to each other through the large grey-shaded shared community, whose semantic contents appear to represent the concept of cross-border air pollution consisting of the international relations between China (中国 *Chūgoku*) and Japan (日本 *Nihon*) concerning the effect (影響 *eikyō*) of PM2.5 (PM) and the need for implementing measures against (対策 *taisaku*) the air pollution (大気汚染 *taiki osen*). Periods five, seven and nine are linked with this large community of words, albeit through fewer and specific nodes, suggesting that the news contents of periods seven, eight or nine cover a narrower topic with specific aspect of PM2.5 compared to the earlier periods. Period two is the richest in content and by far the largest amalgamation of semantic communities. Its contents centre on the health (健康 *kenkō*) risks (リスク *risuku*) and concerns (懸念 *kenen*) of illnesses (疾患 *shikkan*, 肺がん *haigan*), causes (原因 *gen'in*) of pollution such as yellow sand (黄砂 *kōsa*) and spatial origins of air pollution (飛来 *hirai*, 越境 *ekkyō*, 工場 *kōjō*), the roles of the subnational (自治体 *jichitai*) authorities in disclosing (公表 *kōhyō*) and disseminating information (情報 *jōhō*) on their websites (ホームページ homepage) and in raising awareness (呼び掛ける *yobikakeru*) of the risks to the citizens so that they may take precautionary measures by refraining (控える *hikaeru*) from going outdoors (外出 *gaishutu*) when advised. The central word communities in dark grey signify 'health, evaluation, awareness-raising, control, and self-responsible precautionary actions', and they are shared between periods two, three and four. Thus, these communities form the key topics on which the shift of discourse topics between these periods is anchored.

Periods one, two and three are also closely interrelated and characterised by the frequent mention of Beijing (北京 *Peking*) where air pollution occurs (発生 *hassei*) causing respiratory (呼吸 *kokyū*) hazards. Concepts relating to the measurement (測定 *sokutei*) of PM2.5 concentration level (濃度 *nōdo*), small particulate matter (微小 *bishō*, 粒子 *ryūshi*, 物質 *busshitsu*) causing the air pollution (大気汚染 *taikiosen*) from China (中国 *Chūgoku*) act as links between periods two, three and four.

From there the topic of period three develops into a discussion of the results (結果 *kekka*) of study and conditions of relevant illnesses (症状 *shōjō*) as well as the roles of the government (政府 *seifu*) and the subnational areas or regions

(地域 *chiiki*). Periods three and four are mostly interrelated with period two. In other words, the topics of periods three and four derive from period two; they simply took different turns to develop a specific media discourse, each highlighting a distinct aspect of PM2.5 cross-border air pollution.

Linking to periods two and three through the word group of awareness raising (注意 *chūi*, 喚起 *kanki*) and provisional indicator value (暫定指針値 *zantei shishinchi*), period four displays a word group representing details of precautionary measures (屋外 *okugai*, 激しい *hageshii*, 運動 *undō*) that prefectural residents (県内 *kennai*) must take when warnings based on the higher reading (上昇 *jōshō*, 数値 *sūchi*) for air pollution are issued at a specific time of the day (午前 *gozen*, 午後 *gogo*). The transition from period three to four appears to indicate that, by period four, a monitoring and warning system for cross-border air pollution has been put in place, and that the subnational political authorities start to routinely announce the observed readings for pollution.

Period five is partially connected to period two through the word 'this year' (今年 *kotoshi*). The discourse of period five appears to reestablish or normalise the routine of observation and taking necessary measures by referring to the previous year (昨年 *sakunen*). A new actor, corporations (企業 *kigyō*), comes into play in this period, paving the way for the next news topic in period six.

Period six creates a satellite of its own. Unrelated to other periods, its semantic pattern points to a news discourse centering around the current (現在 *genzai*) new (新た *arata*) energy (エネルギー *enerugī*) policy (政策 *seisaku*) relating to product (製品 *seihin*) development (開発 *kaihatsu*), consumers (消費 *shōhi*) and the international (国際 *kokusai*) market (市場 *shijō*), reminiscent of the discourse of the 'Fourth Energy Plan' (*Dai-yonji Enerugī Keikaku*) of the Abe administration. The discourse pattern resonates with the prime minister's speech on Japan's 'adaptation policy' at the UN Climate Summit and it transforms the meaning of 'adaptation policy' from that of the governments measures to protect the citizens from the health hazards of PM2.5 to a whole new discourse of industrial and market opportunity taking advantage (利用 *riyō*) of the pollution.

Each of the last three periods, seven, eight and nine, conveys only a few items, yet they are connected to the largest word group in dark grey. Hence, we know that periods seven to nine share the central theme with periods two to four, but the discourse of the latter revolves around a more specific and defined content. Periods seven and eight are connected with each other through the word 'calling for' (よびかける *yobikakeru*). Semantic contents of period seven signifies issuance of warnings of pollution (発令 *hatsurei*, 発表 *happyō*) to the residents and the contents of period eight indicate that such alarm is communicated by email (メール *mēru*). These periods connect to the largest word group through 'the average value' (平均値 *heikinchi*). Media contents reveal that in these periods high concentration (高い *takai*, 濃度 *kōnōdo*) of PM2.5 was frequently experienced in local cities (市内 *shinai*, 福岡 *Fukuoka*), which was of serious concern (恐れ *osore*) and self-responsible actions were called for.

Period nine is strongly linked to the largest word group (in dark grey), as shown by the thickness of the edges, while its subset is linked to period two through 'small particulate matters' (微小粒子状物質 *bishōryūshijō busshitsu*). Analysis (分析 *bunseki*) and its general use (一般 *ippan*, 使う *tsukau*) may refer to the normalised routine of the warning of dense PM2.5 and responsible measures taken by the public. In fact, what connects period nine to periods three and four is the word 'responsibility' (責任 *sekinin*). We observed earlier that the locus of responsibility of action against the hazards of PM2.5 was associated with the central government and regional authorities in period three, but it shifted to the subnational authorities only in period four, and further shifted to the local residents or general public in periods seven, eight and nine. Moreover, even though the topic of periods seven to nine, which concerns awareness-raising and precautionary measures, resembles the discourse of periods two, they are distinct in orientation. After the intervention of the market-oriented discourse in period six which was produced at the time of COP20, the media discourse of periods seven to nine appear to normalise the system established during periods two to four. That is, the system of issuing announcement on risks of PM2.5 and its triggering of the public's reaction to take self-responsible measures.

To sum up, each of the nine periods exhibits a distinct time-varying pattern, which shifts from one period to the next, sharing, sometimes largely or minimally, certain aspects of the topic of cross-border air pollution. We conclude that hypothesis (b) is supported.

Now that we have identified clearly the shift of discourse patterns between distinct time periods, our next question is how and from what linguistic constituents such patterns are formed. We will next investigate the text-internal structure of interaction between major semantic constituents, traced in prominent word groups, of the media discourse of PM2.5 in our data. To identify the number of clusters that are likely to be representative of the word groups, an annexation plot was drawn as in Figure 5.4.

From the above result, the number of clusters is deemed to be circa five or six and to be reflected in the multivariate analysis shown in Figure 5.5. The contents of each cluster may be summarised as: health concerns (*kenkō*), sources of pollution (*nōdo, hirai*), measures (*taiō*) against the pollution such as control (*kisei*) and scientific analysis (*bunseki*), authorities and policies (*seifu, hōshin*), international such as China (*Chūgoku*), national (Japan: *nihon*), subnational (*jichitai*), societal (*shimin*) and other actors. These point to the types of semantic constituents that may interact with each other to instantiate a particular pattern of discourse.

We deduce from the results that actions, actors and pollution sources are major media contents relating to PM2.5, and how these parameters interact with each other to characterise the media discourse at a given time will be our next object of investigation as expressed in hypothesis (c):

(c) Different media discourse patterns observed at different time periods shed light on how different actors, actions and sources of hazard interact to instantiate that discourse at a given time.

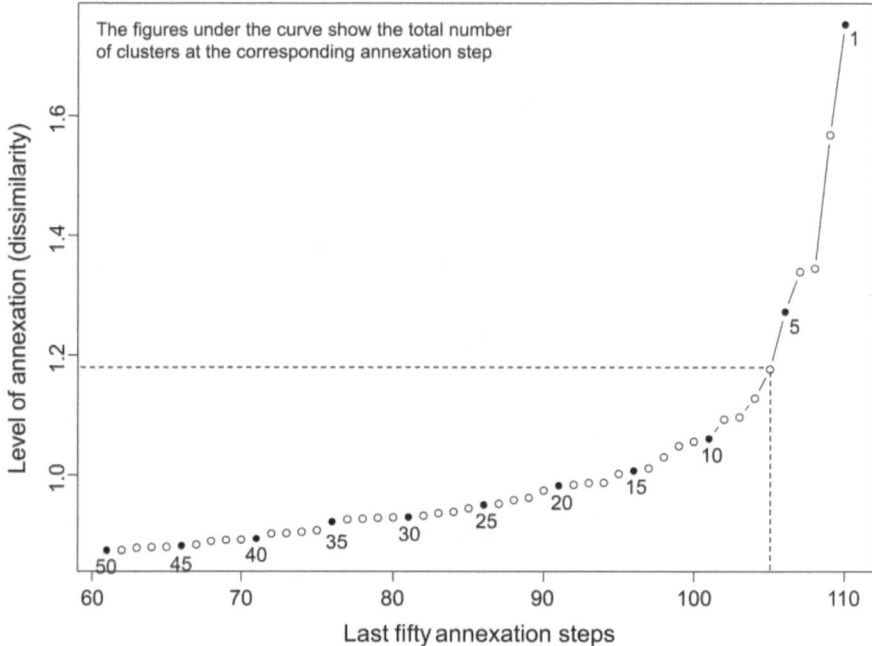

Figure 5.4 Annexation plot
Source: van der Does-Ishikawa (2016)

The hypothesis is tested using 'coding rules' and co-occurrence networks for analysis.

Coding the text-internal relationship of concepts

Using 'coding rules' for analysis, we explore the interactive relationship between the thematic concepts of PM2.5, risk and responsibility in the media discourse of PM2.5, on the one hand, and the concepts of actions, actors and sources, on the other. The coding-rules method allows us to interrogate the internal structure of a text from a specific viewpoint, concept or aspect of discourse by highlighting a particular set of semantic groups and investigating how each interacts with the other sets of words within the same text. Our original coding rules are briefly described below.

Based on our theoretical approach (Hook *et al.* 2015) we posit *Actors*, *Actions*, and information related to PM2.5 (termed *Sources* here for the sake of convenience) as the variables interacting with each other to create a tapestry of political and social discourse at a given time. In this case, 'information' has been divided into the 'news topic' (PM2.5, cross-border air pollution) and the

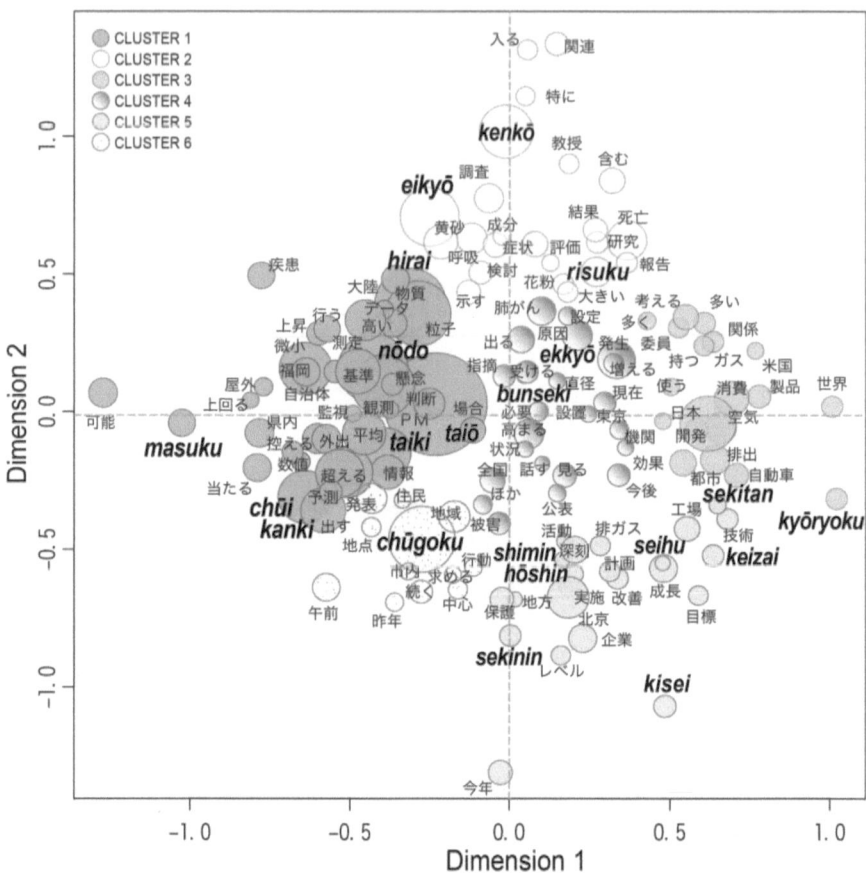

Figure 5.5 Distinct word groups observed in multidimensional scaling (MDS)
Source: van der Does-Ishikawa (2016)

'target concepts of investigation' (risk, responsibility). These are employed as keywords to assemble the data set, on the one hand, and as a set of words labelled 'sources,' on the other. The sources inform the Actors (AR) and Actions (AN) in the discourse being interrogated. AR, AN and Sources (S) are further subcategorised in terms of different conceptual aspects, yielding twenty-one concept groups. Then words found within the top 250 of the most frequently occurring words[7] in the entire data set (PM2.5 + risk + responsibility), and that which occurred in a minimum of 20 texts, are classified into 21 concept groups, as shown in Table 5.3.

Next, the relationships between the external parameters are scrutinised and formulated in a 'heatmap' that is a cross-tabulation between the categories of external variables and the categories of coding rules. The relative values between

Table 5.3 Coding rules and codes for risk and responsibility versus Action, Actor and Sources

Target concepts of investigation	
RISK/RESP-NK	News topics
RISK-GC	Risk-General concepts
RESP-GC	Responsibility-General concepts
Actions	
ACTION-GN	Action-General
ACTION-PL	Action-Policy
ACTION-CT	Action-Controlling
ACTION-AR	Action-Awareness raising
ACTION-CE	Action-Collaborative efforts
ACTION-PR	Action-Precaution
Actors	
ACTOR-IG	Actor-Interest group
ACTOR-EX	Actor-Scientific experts
ACTOR-IN	Actor-Industry/Market
ACTOR-WL	Actor-International/intergovernmental
ACTOR-SA	Actor-Subnational authorities
ACTOR-NA	Actor-National authorities
ACTOR-CZ	Actor-Citizens
Source-related word groups	
SOURCE-OG	Source-Causes-origins
SOURCE-CX	Source-Causes-explanatory
SOURCE-SS	Source-Science-substance
SOURCE-AP	Source-Scientific approach
SOURCE-HM	Source-Health matters

Source: van der Does-Ishikawa (2016)

these coded groups of words are explored and their cross-tabulation is visualised in the heatmap shown in Figure 5.6. The occurrence rates of the words in the above coding scheme under each concept group in the news articles are indicated in percentages.

Two large groups (i.e. clusters) of related concepts are identified. One belongs to the group headed by the newsworthiness of the central theme: PM2.5 and cross-border air pollution (RISK/RESP-NK), and another belongs to the group of general concepts of responsibility and risk (RESP-GC, RISK-GC). The relationships between the subgroups of the former group indicate that news of cross-border air pollution is linked to the explanation of its cause (SOURCE-CX) and substances (SOURCE-SS), its control (ACTION-CT) and the actions of awareness-raising (ACTION-AR) carried out by national (AR-NA) and international (AR-WL) actors.

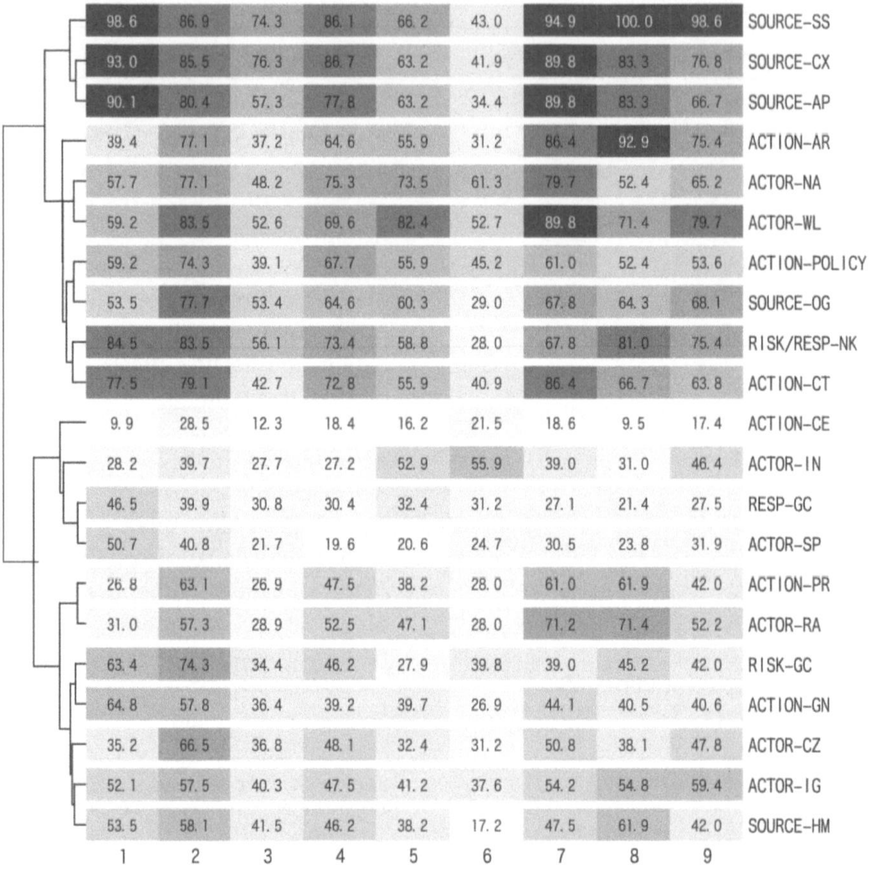

	1	2	3	4	5	6	7	8	9	
	98.6	86.9	74.3	86.1	66.2	43.0	94.9	100.0	98.6	SOURCE–SS
	93.0	85.5	76.3	86.7	63.2	41.9	89.8	83.3	76.8	SOURCE–CX
	90.1	80.4	57.3	77.8	63.2	34.4	89.8	83.3	66.7	SOURCE–AP
	39.4	77.1	37.2	64.6	55.9	31.2	86.4	92.9	75.4	ACTION–AR
	57.7	77.1	48.2	75.3	73.5	61.3	79.7	52.4	65.2	ACTOR–NA
	59.2	83.5	52.6	69.6	82.4	52.7	89.8	71.4	79.7	ACTOR–WL
	59.2	74.3	39.1	67.7	55.9	45.2	61.0	52.4	53.6	ACTION–POLICY
	53.5	77.7	53.4	64.6	60.3	29.0	67.8	64.3	68.1	SOURCE–OG
	84.5	83.5	56.1	73.4	58.8	28.0	67.8	81.0	75.4	RISK/RESP–NK
	77.5	79.1	42.7	72.8	55.9	40.9	86.4	66.7	63.8	ACTION–CT
	9.9	28.5	12.3	18.4	16.2	21.5	18.6	9.5	17.4	ACTION–CE
	28.2	39.7	27.7	27.2	52.9	55.9	39.0	31.0	46.4	ACTOR–IN
	46.5	39.9	30.8	30.4	32.4	31.2	27.1	21.4	27.5	RESP–GC
	50.7	40.8	21.7	19.6	20.6	24.7	30.5	23.8	31.9	ACTOR–SP
	26.8	63.1	26.9	47.5	38.2	28.0	61.0	61.9	42.0	ACTION–PR
	31.0	57.3	28.9	52.5	47.1	28.0	71.2	71.4	52.2	ACTOR–RA
	63.4	74.3	34.4	46.2	27.9	39.8	39.0	45.2	42.0	RISK–GC
	64.8	57.8	36.4	39.2	39.7	26.9	44.1	40.5	40.6	ACTION–GN
	35.2	66.5	36.8	48.1	32.4	31.2	50.8	38.1	47.8	ACTOR–CZ
	52.1	57.5	40.3	47.5	41.2	37.6	54.2	54.8	59.4	ACTOR–IG
	53.5	58.1	41.5	46.2	38.2	17.2	47.5	61.9	42.0	SOURCE–HM

Figure 5.6 Heatmap with the occurrence rates of the members of concept groups
Source: van der Does-Ishikawa (2016)

In the latter group, two large subgroups are identified. In the former, responsibility (RESP-GC) is primarily associated with scientific experts (ACTOR-SP) and industry (ACTOR-IN) as well as collaborative efforts (ACTION-CE). The other group has four subgroups in which risk (RISK-GC) is strongly associated with civilian interest groups (ACTOR-IG), citizens (ACTOR-CZ) with health concerns (SOURCE-HM), and subnational authorities (ACTOR-RA) with precautionary measures (ACTION-PR), and these are closely associated with policy to tackle the PM2.5 hazard (ACTION-POLICY).

In short, the clusters in the heatmap indicate that national and international actors are primarily associated with the scientific study of the hazard and identification of the cause, explanation of the hazard and raising awareness, as well as control of the harmful substance. Subnational and societal actors, on the other hand, are associated with risk and responsibility directly through practical actions of precautionary measures and policy implementation in a concerted effort to gain protection against health risks and hazards, and in case of policy implementation, clearly, the loci of responsibility rest not with international and national actors, but with regional and societal actors.

Next, we will further evaluate the findings presented in Figure 5.6 in the context of the results of co-occurrence network analysis (Figure 5.3) and the historical events described in the section titled 'Historical analysis'.

Scientific explanation of the causes and substances are linked with newsworthiness, and they exhibit the same pattern. They appear to surge and fall with the timing of news related to the occurrence of serious air pollution in China. Actions to draw up and implement precautionary policies in response to the pollution by national and international authorities are strong in periods two and four and periods seven and eight. Responsibility (RESP-GC) emerges as a relatively strong presence in period one, but decreases to circa 30 per cent by period three and continues in the background (Figure 5.6). Risk (RISK-GV) is strongly emphasised during period one and especially in period two, then period four, and again in period eight, and it mirrors the pattern of health matters (SOURCE-HM), indicating the close association between RISK and HEALTH in the media discourse. This coincides with the fact that periods one, two and three concern coverage of the cross-border air pollution from China and the risks to health and potential harms posed, while the news coverage in period six concerns market solutions and international concerted efforts in Japan's adaptation policy (Figure 5.3 and in the section 'Historical analysis'). So far we have identified how the different combinations of concepts contribute to creating distinct media contents in each of the nine periods. Therefore, we conclude that hypothesis (c) is supported.

We also hypothesised that:

(d) Shifts in the pattern of discourse across periods display changes in the association between certain actors, actions and sources of hazards (PM2.5).

The above analysis revealed the evident link between the historical events and the text-internal shifts in media discourse. Emphasis on different aspects of cross-border air pollution in different periods was observed and distinct combinations of actors, actions and 'sources' were found to instantiate different media contents. Hence, we conclude hypothesis (d) is supported.

Finally, the last hypothesis is tested:

(e) Shifting discourse patterns on risk and responsibility over time display a phenomenon of top-down 'off-loading' of responsibility from the state to other actors coupled with the channeling of momentum into individual actions based on 'self-responsibility'.

A co-occurrence network analysis is employed in combination with coding rules below to identify how the general concepts of risk (RISK-GC) and responsibility

(RESP-GC) and other parameters interact with each other in each of the nine periods, and how the combinations of specific parameters may instantiate discourse assigning the locus of responsibility to a specific actor concerning a particular 'responsible' action. The results will be compared to the earlier statistical results and contextualised for interdisciplinary evaluation using the techniques of the historical branch of critical discourse analysis.

The relationship between the parameters of ACTION, ACTOR and SOURCE and the temporal variable (i.e. the nine periods) is demonstrated in Figure 5.7, elucidating which of the combinations of these parameters are particularly prominent at a given period. We will now consider the shifting loci of responsibilities by applying multifaceted, multidisciplinary analyses that combine the results of co-occurrence network of frequently occurring words per period (Figure 5.3), cross-tabulation of the occurrence rates of different parameters (Table 5.4),

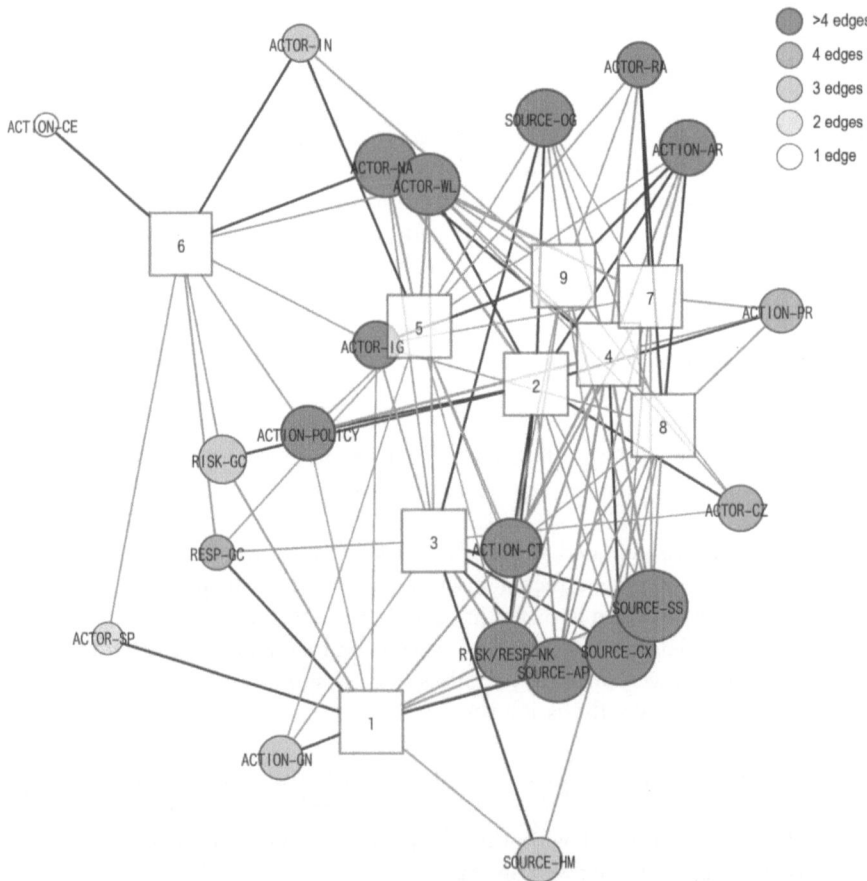

Figure 5.7 Co-occurrence network of coding rules and the nine periods
Source: van der Does-Ishikawa (2016)

shifting weight of parameters over time (Figure 5.6) and qualitative analyses of historical events.

The temporal shift in the locus of responsibility

The relationship between the concepts of RISK (risk) and RESP (responsibility) and their intra- and intertextual relation to the concepts of Actors, Actions and Sources are explored using a 'co-occurrence network'. The network plots 'connect words' co-occurring in a text with lines (*edges*) that vary in thicknesses according to the strength of the relationship between the words. The plot identifies centrality, too; namely, the degrees of importance accorded a particular word in the text when functioning as a node in the network structure in the given instance of discourse. The parameters are set as follows: the minimum word (i.e. POS) occurrence is 35, the minimum number of articles in which a target word occurs is 25 and the number of POS for analysis is 382. Under these criteria, the number of *edges* plotted was 60 ± 15, adjusted according to the density of *edges* in each case.

The previous result of co-occurrence network analysis (Figure 5.3) showed that risk in period one was associated with Japan's past experience of pollution or international reports on the potential harmful impact of PM2.5 on health. Here, period one is associated with the concept of general responsibility that connects directly and strongly with high centrality. Weak connection to the general concept of risk and strong connection with the concept of general action indicates a news discourse in which the 'risks' of pollution were still undefined, but it had already triggered the concept of 'responsibility'. Specialists are the only prominent actors involved in this period. Using the techniques of critical discourse analysis, a brief historical contextualisation will prove effective here.

The potential harms caused by PM2.5 pollution have been well known in Japan since the 1970s, but comprehensive precautionary measures taken in response to the risks posed were not implemented at the time, unlike in the case of suspended particulate matter (SPM), for which subnational political authorities are obliged to issue warnings and to impose emission reduction measures. It was only in September 2009 that environmental standards relevant to PM2.5 were established under the Fundamental Environment Act, Article 16(1) in ministerial announcement No. 33 on the Environmental Standards Relating to Air Pollution caused by Microparticulate Matter (*biryūshi-jō busshitsu ni yoru taiki-osen ni kakaru kankyō-kijun ni tsuite*). The Ministry of Environment issued Administrative Procedure Standards regarding Continuous Surveillance of Atmospheric Pollution Based on the Provisions of Article 22 of the Air Pollution Control Act (*taiki-osen Bōshi-hō Dai-22-jō no Kitei ni motozuku Taiki no Osen no Jōkyō no Jōji-kanshi ni kansuru Jimu no Shori-kijun ni tsuite*) in March 2010. In the same year the ministry revised the Manual for Continuous Surveillance of Ambient Air Pollution (*Kankyō Taiki Jōji Kanshi Manuaru*) and designated constant monitoring of PM2.5 concentration as a statutory responsibility entrusted to subnational political authorities (prefectural and city governments). Then, in July 2011, Guidelines for Atmospheric Composition Analysis (*Seibun bunseki gaidolain*) were issued, which were followed in April 2012 by the Atmospheric Composition Measurement Manual (*Seibun-bunseki manuaru*). In this context the result of the co-occurrence network underscores the fact that media discourse in period one revolves around the emerging new risk posed by PM2.5 inviting reference by professional and scientific experts to past experience of pollution to make sense of the risk, while offering the undefined locus of responsibility to tackle the new threat.

In period two (February 2013 to July 2013) the newsworthiness of the concepts of risk and responsibility is linked to PM2.5 most strongly. This period came immediately after the incidents of the week-long severe air pollution in Beijing in mid to late January 2013. The Beijing Municipal Environmental Monitoring Center announced that some of their monitoring stations had recorded PM2.5 concentration levels of over 700 µg/m³, well above the 'hazardous' level of 500 µg/m³ according to the Revised Air Quality Standards for Particle Pollution and Updates to the Air Quality Index of 14 December 2012. Beijing was thereby prompted to release an orange level haze warning, while adopting its emergency response plan for extreme air pollution. The news was widely reported in Japan and worldwide, prompting a sudden spike of interest in the news coverage of risks posed by PM2.5 and the responsibility of an undefined body (actor) to respond to its potential hazard to health.

Further, a variety of concepts are strongly linked to this period. Together with the heightened newsworthiness of Risk and Responsibility (RISK/RESP-NK) the general concept of Risk (Risk-GC) was reintroduced in the news discourse. Actions associated with this period are awareness-raising (AN-AR), policy (AN-POLICY) making, controlling (AN-CT) the hazard, and precautionary measures (AN-PR). The origins of the pollution (S-OG) were sought. International actors (AC-WL) and the citizens (AC-CZ) affected by the pollution are involved. Judging from these combination of parameters, the news discourse in period two is predominantly characterised by awareness-raising of the risks and negotiating the locus of risk and responsibility.

In periods two and three, the risks of cross-border air pollution are reported establishing China as the source of the pollution. The information communicated, both scientific and social, is dense. Period three is characterised by the scientific effort to explain the pollutants (SS-SS) to fulfil the general responsibility of the national (AR-NL) and international actors (AR-WL) by controlling the hazard (AN-CT). The origins of the pollution (S-OG) are studied based on a scientific approach (SS-AP) in order to explain the causes (S-CX) and its health impact (SS-HM). Investigative reporting on the health risks posed by PM2.5 dominates the media discourse. General responsibility is no longer linked to the national authority (AR-NL), because by that time, it had shifted from the national authority to the sources of pollution or to their scientific explanation.

However, issues of scientific investigation were settled by period four except the need for continuous explanation of the cause of pollution (SS-CX). Subnational actors (AR-RA) and citizens (AR-CZ) are involved in implementing precautionary measures (AN-PR) against the environmental risk under the directions of the national authority (AR-NL). General responsibility (RESP-GC) is not closely linked to any actors in period four, but responsible action (AN-PR) is. This suggests that a general search for the responsible actor to implement an action is no longer a major topic of discourse then, because the system – of the local authorities' alerting the citizens to a risk and them taking precautionary measures in response – had normalised by this period. Consequently, the locus of responsibility had shifted from the national authorities' scientific investigation of pollutants to the subnational authorities and the citizens to implement precautionary measures, and that was also naturalised.

Period five is weakly linked to the concept of general responsibility as well as the newsworthiness of risk and responsibility again, suggesting a change of topic from the previous periods by the introduction of a new actor, industry/market (AR-IN). Continuing from period two national (AR-NL) and international (AR-WL) actors are in the background. Implementing a policy (AN-PO) and raising awareness (AN-AR) concerning the origins of the hazard (SS-OG) are associated with subnational actors (AR-RA), while continuing from period three a scientific approach (SS-AP) to control (AN-CT) the pollutants is a newsworthy topic of risk and responsibility (R-R/NK) and they are closely associated with the industrial actors to whom the general responsibility is assigned.

In period six media discourse moves radically away from the contents of the previous periods, at least on the surface. The concepts of risk and responsibility are redefined by collaborative efforts (AN-CE) led by national and industrial actors aided by specialists (AR-SP) to implement policies. Yet, in fact, the involvement of industrial actors actually began in period five and continues to the next three periods in the background.

Recalling the heatmap (Figure 5.6) there were three waves in the general newsworthiness of cross-border air pollution: periods one to three, four to six, and seven to nine, of which period six stands out in its news contents. National and industrial actors are particularly prominent and such a pattern is quite distinct from any other periods. This period follows from Prime Minister Abe's speech at the UN Climate Summit at the end of period five, where he framed Japan's 'adaptation policy' through 'technological innovation' and energy efficiency. At the beginning of period six, a new discourse of adaptation policy by industrial innovation became prominent as exemplified in the following excerpt:

> 原発推進の背景には、石炭火力に依存するエネルギー需給構造の調整、ＰＭ２・５など大気汚染深刻化、電力需要の拡大などの理由があり、これらの解決には原子力しかないというのが中国の認識。また、地方財政における大きな収入源となっている点なども挙げられる。さらには原子力産業をテコに国際原発市場に参入し、世界の原子力計画の受注シェアを拡大する方針。中国核工業集団公司と中国広核集団有限公司を中心に、すでにパキスタンに原子炉を輸出した実績がある（中略）。

China is promoting nuclear energy for a variety of reasons, among them the need to adjust China's energy supply structure by changing its dependency on coal-fired thermal power; the ever intensifying atmospheric pollution caused by PM2.5 and other pollutants; and the expanding demand for electricity supply. As far as China is concerned, nuclear energy is the only solution to all these pressing issues. Again, nuclear energy has become a major source of income for subnational governments. What is more, China's policy is to use the nuclear industry as a way to join the international nuclear power market and expand its worldwide share of international orders for nuclear plant construction. Indeed, China already has a proven track record in exporting nuclear reactors to Pakistan, headed by the China National Nuclear Corporation (CNNC:中国核工業集団公司) and the China General Nuclear Power Group (CGN: 中国広核集団有限公司).

(Kaku 2014 in *The Nihon Genshiryoku Sangyō Shimbun*, 24 April 2014: 4)

The periods seven and eight see the revived importance of precautionary measures (AN-PR) carried out by the subnational authorities (AR-RA) and citizens (AN-CZ) as reflected in the text-internal structure. Industrial actors (AR-IN) are in the background. Identifying the origin of pollution (SS-OG) by the local authorities (AR-RA) and the taking of precautionary measures (AN-PR) by the citizens (AR-CZ) are central in the discourse of the periods seven to nine signifying a revival of period two, but in a much subdued, normalised form. Linking to period five the concept of responsibility is embedded deeper in the discourse of period nine, whereby the citizens' self-responsibility to take precautionary measures, with the aid of industrial and local actors, had become naturalised.

Turning to the texts of news articles during periods seven to nine, we find that the media at the time were primarily concerned with reports of China's (lack of) environmental efforts and progress and industrial efforts to create environmental 'adaptation' products and even showcasing such products. For example, during period seven, on 24 April 2015, China's first revision of the Environmental Protection Act in twenty-five years was reported, and the opening of the IE2015 International Environment Exhibition in Shanghai on 6–8 May was announced. In period eight, on 15 August 2015, China's new legal framework to punish offenders to cap coal consumption was reported, and on 20 October of the same year a joint research plan between the cities of Tangsha, Hebei in China and Kita-kyushu in Japan was announced. In period nine, 7–8 December 2015, COP21 in Paris was reported, and on 8 December the incident of China's first issuance of a 'Red Alert' in Beijing was extensively covered. News on 5 March 2016 welcomed Prime Minister Li Keqiang's declaration to curb pollution made at China's National People's Congress.

Throughout these periods, reports on monitoring and alerts for a higher concentration of PM2.5 are issued periodically, often accompanied by advertisements on anti-PM2.5 merchandise such as facial masks and air conditioning devices, among other products. News discourse began to present atmospheric pollution as if it is a given, annual or seasonal event. That is, the risk and hazard posed by PM2.5 has become normalised and incorporated into a range of air pollution types such as pollen. In this way, the tripartite steps of scientific observation, issuance of warning by subnational authorities and the adoption of precautionary measures taken by the citizens to self-protect have become routinised. A typical news article below states:

> 春になると、ここ数年必ずニュースになるのが大気を汚す物質「PM2.5」の濃度上昇だ。中国ではPM2.5が濃い霧のようになり、飛行機の離着陸にも支障が出て、外出もできない状態になる。（中略）日本列島では上空の偏西風に乗って高気圧が大陸から移動してくる3〜5月に濃度が上がりやすい。

> For the past few years spring has brought regular reports on the higher concentration of the air pollutant called 'PM2.5'. PM2.5 forms a dense fog in China, affecting take-off and landing of aeroplanes, while preventing people from venturing outside . . . The high concentration of PM2.5 tends to hit Japan between March and May when, carried by the upper westerlies, high atmospheric pressure moves from the Chinese continent to Japan.

> (*The Nihon Keizai Shimbun*, 4 March 2016)

The co-occurrence network representing the nine periods reveals that:

- The state's discourse of responsibility is limited to the scientific observation and explanation of the hazardous substance and identification of the source of pollution for the sake of setting up the observation and warning system nationwide and the raising of awareness among the citizens.
- Actual observation and issuance of warnings are executed by the subnational authorities to implement the policy, and this system has been firmly established by period four.
- However, these authorities do not take further action to protect the citizens from harm.
- By periods five and six industrial actors take centre stage in the policy implementation, backed by national actors.
- Industrial actors as well as societal actors are linked through word groups of 'products' and 'consumer' to implement 'policy'.
- By period four, it is the citizens themselves who are associated most strongly with carrying out self-protective measures in their everyday lives, and this was normalised firmly, with the aid of market actors, by period nine.

What we presented above, then, is concrete evidence of the shifting loci of responsibility to protect the citizens from PM2.5. That is, responsibility cascades down across the nine periods from the national to the subnational level and then on to the market (industry) and societal actors. Hence, we conclude that hypothesis (e) is supported.

Conclusion

This study has deployed both quantitative and qualitative methods to interrogate the shifting locus of responsibility in the media discourse on cross-border atmospheric pollution in Japan. The findings demonstrate how, in responding to the risks and potential harms posed by cross-border PM2.5, responsibility is being off-loaded from the state to subnational political authorities, the market and society, a shift documented in other areas of life (Hook and Takeda 2007).

While a semantic group of the concept of 'risk' tends to be directly expressed in the media discourse, the concept of 'responsibility' is embedded deeper in the tendrils of discourse construction, lacking the salience of 'risk'. Nevertheless, as we mined deeper into the text-internal structure of the discourse, our methods enabled us to excavate words relating to 'responsibility' in the news media reporting cross-border air pollution, as seen in the links to the semantic groups of actors, actions and information sources. As shown, these are in different combinations in different historical time frames between August 1998 and March 2016. The interaction between these groups of concepts shapes the discourse of the news articles in the nine periods covered, and the media discourse of each period presents distinctive characteristics.

We have considered the temporal variation patterns to emerge from the trends in the shifting landscape of risk and responsibility. Period one, before the media captured the readers' attention by covering the high atmospheric pollution in Beijing at the beginning of 2013, focused on scientific information relating to general air pollution and actions taken in response by international and domestic actors. Direct mention of 'responsibility' was the highest in period one. It fades towards period three, picks up in period four, and then decreases again in period seven to the lowest level and remains below 30 per cent of the coverage (Figure 5.6). Meanwhile, the mention of 'risk' begins high at 63.4 per cent in period one and peaks in period two and then drops to 34.4 per cent in period three, before picking up again in period four and then decreasing in period five. It then once again increases in period six. Since the 'responsibility' embedded in period one mostly belongs to the discourse of the past incidents of Japanese industrial pollution, we conclude that the trends of risk and responsibility in the time frame of periods two to nine coincide. If that is the case, then roughly three waves of risk communication and responsibility can be observed in this time frame (periods two and three; periods four, five and six; and periods seven, eight and nine). The first wave swirls around the risks and potential harms posed to health; the second wave concerns the 'international dimension' of Japan's 'adaptation policy'. The co-occurrence network of frequently occurring words testifies to how the contents of period six news texts differ radically from the previous periods, even though the same terminology is in evidence. Discourse patterns of periods five and six are highly associated with national and industrial actors who take the centre stage in the 'adaptation policy' to tackle climate change. The third wave brings back the three-step discourse of risk, vigilance, and self-responsibility that was established in periods two to four, but this time normalising and routinising it. The hazard of PM2.5 is given and the onus of tackling the issues of potential harm is placed on the shoulders of societal actors. The responsibility has cascaded down from the international and national actors to the subnational actors to the industry and finally to the citizens.

The last point alerts us to the possibility of a particularly intriguing finding: namely, that Prime Minister Abe's 'adaptation policy in response to climate change' deploys the selfsame environmental terminology as in previous periods, but the discourse contents betray that, under his administration, Japan's adaptation policy is focusing increasingly on the market (economic revitalisation and industry) rather than on concrete political measures to protect the population's health against the risks and potential harms from climate change. It is as if, in advancing the restructuring of the economy by surfing the wave of neo-liberal globalisation, so-called Abenomics (Miyazawa and Yamada 2015), the prime minister baulks at accepting the state's responsibility to tackle directly PM2.5. Herein lies a clearly identifiable link between state policy on cross-border pollution from China and the everyday health of the nation. Clearly, what is needed is not only a policy of off-loading the state's responsibility to actors in the market and society, but one where the state, too, shoulders the responsibility to protect the nation's health as well as respond to the risks and potential harms to the citizens arising from climate change.

Appendix

Table 5.4 English translation of coded words in alphabetical order

アジア	Ajia	Asia
アレルギー	arerugī	allergy
ばい煙	bai'en	soot and smoke
米国	beikoku	US, United States of America
微（小）粒子	bi(shō)ryūshi	micro-particles
防止	bōshi	prevention, prevent, deter, inhibit, protect, deterrent
分析	bunseki	analyse, analysis
分担	buntan	assignation, burden, share, division
物質	busshitsu	substance
病院	byūin	hospital
値	chi/atai	value, figure
地方	chihō	local, regional, area, provincial, locality
地域	chiiki	region, regional
知見	chiken	knowledge, finding, perception
知識	chishiki	knowledge, idea, experience, education, comprehension, cognition
窒素	Chisso	nitrogen, azoto
調査	chōsa	survey, investigation, research
中国	chūgoku	China, Chinese
注意	chūi	caution, attention
大学	daigaku	university
大臣	daijin	minister
団体	dantai	association, community, group, society, body
データ	dēta	data
影響	eikyō	effect, influence
ディーゼル	dīzeru	diesel
疫学	ekigaku	epidemiology
越境	ekkyō	cross-border

エネルギー	enerugī	energy
NGO	enu-jī-ō	NGO
含む	fukumu	include, involve
福岡	Fukuoka	Fukuoka
外務省	gaimushō	Ministry of Foreign Affairs
外出	gaishutsu	going out
学会	gakkai	scientific meeting, society, workshop, academic meeting, learned society, academic conference
学校	gakkō	school
ガス	gasu	Gas
原因	gen'in	cause, causality, reason, explanation, ground, breeder
原子力	genshiryoku	nuclear power, atomic energy, nuke, atom
現在	genzai	at present, right now, at present day
義務	gimu	obliation, burden, duty, imposition, liability, onus, responsibility, commitment
合意	gōi	accord, agree, agreement, arrangement, settlement, understanding, treaty
行政	gyōsei	public administration, state, executive branch, government
激しい	hageshii	active, acute, stark, violent, wild, intense
肺	hai	lung, pulmonary
排出	haishutsu	issue, announce
販売	hanbai	sales, sell, sale, distribute, vendition, distribution
判断	handan	judge, judgement, assess, assessment, verdict
反応	hannō	response
発表	happyō	announcement, presentation
発生	hassei	incidence, outset, generation, accrual
発症	hasshō	onset (illness), occur (events), pathogenesis
果す	hatasu	accomplish, perform, achieve
発令	hatsurei	issue, announce, announcement
平均	heikin	average, mean
被害	higai	damage, injury, harm
控える	hikaeru	refraining, refrain, sparing, abstaining, abstain, hold back
飛来	hirai	fly over, inbound
人	hito	person, human-being, mankind, people

(*Continued*)

Table 5.4 (Continued)

必要	hitsuyō	necessary, necessity, needs, need
報道	hōdō	report, news
保健	hoken	health
ホームページ	hōmupēji	homepage
評価	hyōka	evaluate, evaluation
委員	iin	a committee member, commissioner
一般	ippan	general, the general, the public
医師	ishi	medical doctor, physician
実施	jicchi	implement, implementation
自治体	jichitai	municipality, autonomy, local town
児童	jidō	pupil
自動車	jidōsha	automobile
自己	jiko	ego, self
自主	jishu	autonomy, freedom, self-, independent, voluntary
自主規制	jishu-kisei	self-regulate, voluntary restraint, voluntary restriction, self-imposed curb
自粛	jishuku	self-control, self-censorship, voluntary ban
情報	jōhō	information, news
条件	jōken	conditions, provision
上昇	jōshō	rise, ascend, climb, elevation, elevate, ascent, ascension
住民	jūmin	residents
循環器	junkanki	circulatory organ
課題	kadai	challenge, task
花粉	kafun	pollen
科学	kagaku	science, scientific
海外	kaigai	overseas
開発	kaihatsu	development
回避	kaihi	avoidance, avoid, edlusion, elude, dodgery, dodge, manoeuvre around, fence off, shun, weasel
解明	kaimei	clarification, clarify, elucidation, elucidate, interpretation, interpret, resolution, breakthrough
会社	kaisha	company, business, firm, office, shop
改善	klaizen	improve, improvement, upgrade, remediate, ameliorate, amelioration
患者	kanja	patient
喚起	kanki	raise, raising, evoke, evoking, arouse, resonate
韓国	kankoku	Korea
環境相	kankōsō	environment minister
環境汚染	kankyō-osen	environmental pollution

環境省	kankyōshō	MoE, Ministry of Environment
関連	kanren	relation, relate, concern, relevant, relevance
管理	kanri	manage, administer, control, direct, govern, governance, administration
監視	kanshi	monitor, monitoring, observation, surveillance, watch
関心	kanshin	interest, heed, notice, regard, respect for
観測	kansoku	observe observation
監督	kantoku	direct, manage, director, manager, supervise, surveillance, take charge of
家庭	katei	home, family
活動	katsudō	activity
経営	kei'ei	management, manage, administer
警報	Keihō	warning, alert, alarm
警戒	keikai	alarm, caution, precaution, vigilance, watchout
計画	keikaku	plan, design, device, project, schedule, program
傾向	keikō	trend, tendency
経済	keizai	economy, economics, economic
結果	kekka	results, outcome
県	ken	prefecture, prefectural
県民	ken'min	residents of a prefecture
県内	ken'nai	inside the prefecture
懸念	Kenen	concern, fear
健康	kenkō	health
研究	kenkyū	research, study, investigate, investigation, survey
研究所	kenkyūjo	research institution
県内	kennai	residents of the prefecture
検討	kentō	review, consider, consideration, inquest
気をつける	ki-wo-tsukeru	take care
企業	kigyō	business, company, corporate enterprise, industry, corporation, holding
基準	kijun	standards
気管支	kikanshi	bronchi, bronchus
危険	kiken	danger
気候変動	kikō-hendō	climate change
禁煙	kin'en	no smoking, quitting smoking, a ban on smoking, abstention from tobacco
記録	kiroku	record
規制	kisei	regulation, control, regulatory

(*Continued*)

Table 5.4 (Continued)

気象	kishō	meteorological phenomena, weather
空気	kūki	air
子供	kodomo	child, children, young, youngster
超える	koeru	exceed
個人	kojin	individual, individuality, an individual person
工場	kōjō	plant, factory, manufacturing building, industrial establishment
効果	kōka	effect, impact, advantage, outcome, result
光化学	kōkagaku	photochemistry, actinochemistry
協会	kōkai	association
国民	kokumin	citizen, citizenry, nation, people, public
国内	kokunai	domestic
国際	kokusai	international
呼吸	kokyū	respiration, breathing, respiratory, wind
高齢	kōrei	elderly, older, senior citizen
考慮	kōryo	consider, take into account
協力	kōryoku	concerted effort, alignment, align, tandem, team-up, cooperation, cooperate, collaboration, collaborate
黄砂	kōsa	yellow sand, yellow dust
国	kuni	country, state, national, nation
今年	kotoshi	this year
協調	kyōchō	collaborate, collaboration, cooperate, cooperation, harmonize, harmonization, partnership
共同声明	kyōdō-seimei	joint statement, announcement, declaration, communique
強化	kyōka	strengthen
拠点	kyoten	base, position, stronghold, hub, outpost, lodgement
九州	Kyūshū	Kyūshū
マスク	masuku	face mask
認める	mitomeru	accept, admit, recognize, affrim, allow, credit, confess, endure, let on, license, permit, sanction, seal, come round
メール	mēru	email
問題	mondai	issue, problem
基づく	motozuku	based on
持つ	motsu	have, hold, carry, keep
向ける	mukeru	direct, towards, face
日中韓	nic-chū-kan	Japan, China and Korea

日本	Nihon	Japan
担う	ninau	assume, bear, take up, play a role in
任務	ninmu	mission, role, task, assignment, commission, duty
濃度	nōdo	concentration (level)
行う	okonau	act, do, proceed, implement
起こす	okosu	start, bring, raise, institute, operate, set up
屋外	okugai	outdoors, out of doors
思う	omou	think, consider, feel
汚染	osen	pollution
押し付ける	oshitsukeru	foist, lay, enforce, push, shirk, lay down, work off, obtrude
欧州	ōshū	Europe
恐れ	osore	fear, risk
北京	Pekin	Beijing
PM2.5	pīemu-ni-ten-go	PM2.5
連携	renkei	cooperation, cooperate, tandem, tie, linkage
リスク	risuku	risk
定める	sadameru	set up, appoint, meet
削減	sakugen	reduction, curtailment, cut, cutback, reduce, cutdown, slash, rundown
産業	sangyō	industries, trade
設置	secchi	install, installation, set
参加	sanka	participate, take part in, attend
せい	sei	attributable, responsible for, ascribe
整備	seibi	provide, consolidate, maintain, condition
成分	seibun	composition
政府	seifu	government
製品	seihin	product, finished goods, manufactured article
生活	seikatsu	life, living
政策	seisaku	policy
生産	seisan	production, produce, manufacture
生成	seisei	generate, prepare
世界	sekai	world, globe
責任	sekinin	responsibility, answerability, obligation, onus, charge, commitment, blame, burden
石炭	sekitan	coal
専門	sen'mon	expert, expertise, specialist, speciality, professional, specialised
戦略	senryaku	strategy, strategic
センター	sentā	centre

(*Continued*)

Table 5.4 (Continued)

説明	setsumei	explain, explanation, account for, commentary, clarification, directions elucidation, description, review, report, representation
設定	settei	parameters, configuration, set-up
市	shi	city
死亡	shibō	death, decease, deceased
支援	shien	support, backup, help, patronage, alignment
市場	shijō	market
疾患	shikkan	disease, illness, malady, complaint, condition
市民	shimin	residents of the city
市内	shinai	inside the city
深刻	shinkoku	serious, seriousness, profound, significant
指針	shishin	guidelines, guiding principle, index, principle, policy, active pointer
指摘	shiteki	point out, point, bring up, suggest
省	shō	department, ministry
処置	shochi	treat, treatment, procedure, take action
消費者	shōhi-sha	consumers
消費量	shōhiryō	volume of consumption
症状	shōjō	symptom, case, condition, indication, complaint
装置	sōchi	equipment
速報	sokuhō	flash report, preliminary report, advance announcement, breaking news
測定	sokutei	measuring, measurement
存在	sonzai	existence, being, entity, occurrence
訴訟	soshō	action, case, cause
大気	taiki	atmosphere, atmospheric air, ambient air
大気汚染	taiki-osen	air pollution, atmospheric pollution
大国	taikoku	big power, great power, major nation, world power
対応	taiō	respond
大陸	tairiku	continent, mainland
対策	taisaku	measures (against)
体制	taisei	system
態勢	taisei	arrangement, preparations, set-up
高い	takai	high
達成	tassei	achieve

転嫁	tenka	imputation, impute, shift, redirect, dodge, pass on
都道府県	todōfuken	prefectural and city governments
東京	Tōkyō	Tokyo
当局	tōkyoku	authority
都市	toshi	city, urban
問う	tou	ask, question, inquire, bring/press (charge)
追求	tsuikyū	pursue, pursuit, pursuance, search, follow-up
受ける	ukeru	suffer, receive, given
使う	tsukau	use, employ, deploy
運動	undō	activity (physical or else), action, campaign, drive, motion, move(ment)
訴え	uttae	plead, pleading, accuse, complain, accusation
枠組み	wakugumi	framework
負う	ou	bear, undertake, assume, bear, incur
呼び掛ける	yobikakeru	address, urge, call on, call for
要因	yōin	cause, contributing factor, element, trigger
予想	yosō	predict, prediction, expect
予測	yosoku	forecast, predict, prediction, project, projection, prognosis
有機化合物	yūki-kagōbutsu	organic compound
暫定	zantei	interim, provisional, transitional, temporary
全域	zen'iki	entire area, whole area
全国	zenkoku	nationwide, national, nationally, whole of country
喘息	zensoku	asthma

* English equivalents are not exhaustive and for reference only.

Source: van der Does-Ishikawa (2016)

Notes

1 This aspect is dealt with in greater detail in a separate paper (see Hook and van der Does-Ishikawa forthcoming).
2 Original in Japanese: これまでの研究成果を国民の各界各層に広く発信し、地域・自治体における気候変動影響と適応研究、及び適応策の一層の促進に資するため。
3 Examples of 'self-responsible' actions include citizens carrying portable, handheld devices for keeping track of their own exposure to air pollutants (Penza *et al.* 2014) or regulating indoor air quality in 'smart cities' (Curry 2016).
4 Personal communication with veteran journalists at the *Chūgoku Shimbun*.

5 The authors thank Professor Higuchi Kōichi at Ritsumeikan University and Dr Suzuki Takashi of TSC Consulting for their expert advice on this new multidisciplinary approach to text analysis.
6 The article from 4 August 1998 was the oldest data we retrieved from the database search using the keywords PM2.5, risk and responsibility.
7 Content words only. The number of words subject to coding rules analysis (250 in this case) was determined by the frequency of occurrences: the number of texts in which the word occurred against the total number of texts in the database.

References

Abe, Shinzo (2014) *Statement by Prime Minister Shinzo Abe at the Plenary Session of the UN Climate Summit 2014*. Delivered on 23 September 2014. Accessed on 14 July 2016 from www.mofa.go.jp/ic/ch/page3e_000235.html

Curry, David (22 April 2016) 'Could smart cities improve indoor air quality with home sensors?'. *Readwrite*. Accessed on 6 October 2016 from http://readwrite.com/2016/04/22/sensors-improve-indoor-air-quality-ct4/

Denton, Fatima, Thomas J. Wilbanks, Achala C. Abeysinghe, Ian Burton, Quingzhu Gao, Maria Carmen Lemos, Toshihiko Masui, Karen L. O'Brien, and Koko Warner (2014) 'Climate-resilient pathways: Adaptation, mitigation, and sustainable development', in C.B. Field, V.R. Barros, D.J. Dokken, K.J. Mach, M.D. Mastrandrea, T.E. Bilir, M. Chatterjee, K.L. Ebi, Y.O. Estrada, R.C. Genova, B. Girma, E.S. Kissel, A.N. Levy, S. MacCracken, P.R. Mastrandrea, and L.L. White (eds) *Climate Change 2014: Impacts, Adaptation, and Vulnerability. Part A: Global and Sectoral Aspects. Contribution of Working Group II to the Fifth Assessment Report of the Intergovernmental Panel on Climate Change*. Cambridge, UK, and New York: Cambridge University Press, pp. 1101–1131. Accessed on 27 July 2016 from http://ipcc-wg2.gov/AR5/images/uploads/WGIIAR5-Chap20_FINAL.pdf

Dershowitz, Alan M. (2007) *Preemption: A Knife that Cuts Both Ways (Issues of Our Time)*. New York: Norton and Company.

de Swert, Knut (2012) *Calculating Inter-Coder Reliability in Media Content Analysis Using Krippendorff's Alpha*. Accessed on 20 July 2015 from www.polcomm.org/wp-content/uploads/ICR01022012.pdf

Edenhofer, Ottmar, Ramón Pichs-Madruga, Youba Sokona, Jan C. Minx, Ellie Farahani, Susanne Kadner, Kristin Seyboth, Anna Adler, Ina Baum, Steffen Brunner, Patrick Eickemeier, Benjamin Kriemann, Jussi Savolainen, Steffen Schlömer, Christoph von Stechow, and Timm Zwickel (eds) (2014) *Climate Change 2014 Mitigation of Climate Change: Working Group III Contribution to the Fifth Assessment Report of the Intergovernmental Panel on Climate Change*. Cambridge, UK and New York: Cambridge University Press. Accessed on 27 July from www.ipcc.ch/pdf/assessment-report/ar5/wg3/ipcc_wg3_ar5_full.pdf

Fairclough, Norman (1989) *Language and Power*. London: Longman.

Franklin, Meredith, Ariana Zeka, and Joel Schwartz (2007) 'Association between PM2.5 and all-cause and specific-cause mortality in 27 US communities'. *Journal of Exposure Science and Environmental Epidemiology* 17: 279–287.

Hamra, Ghassan B., Neela Guha, Aaron Cohen, Francine Laden, Ole Raaschou-Nielsen, Jonathan M. Samet, Paolo Vineis, Francesco Forastiere, Paulo Salvida, Takashi Yorifuji, and Dana Loomis (2014) 'Outdoor particulate matter exposure and lung cancer: A systematic review and meta-analysis'. *Environmental Health Perspectives* 122(9): 906–912.

Harrison, Roy M., D.J.T. Smith, and A.J. Kibble (2004) 'What is responsible for the carcinogenicity of PM2.5?'. *Occupational and Environmental Medicine* 61: 799–805.

Higuchi, Kōichi (2004) 'Tekisuto gata dēta no Keiryōteki-bunseki, Futatsu no apurōchi no Shunbetsu to Tōgō, Riron to Hōhō'. *Sūri-shakai Gakkai* 19(1): 101–115.

Hill, Christopher and Karen E. Smith (2000) *European Foreign Policy: Key Documents in Association with the Secretariat of the European Parliament.* New York: Routledge.

Hook, Glenn D. (2012) 'Recalibrating risk and governing the Japanese population'. *Critical Asian Studies* 44(2): 309–327.

Hook, Glenn D. and Luli van der Does-Ishikawa (forthcoming) 'The governmental health risk-management to citizen's self-responsive actions: Cross-border Atmospheric Pollution in Japan'.

Hook, Glenn D., Ra Mason, and Paul O'Shea (2015) *Regional Risk and Security in Japan: Whither the Everyday.* Sheffield Center for Japanese Studies/Routledge Series. London: Routledge Taylor & Francis Group.

Hook, Glenn D. and Hiroko Takeda (2007) '"Self-responsibility" and the nature of the postwar Japanese state: Risk through the looking glass'. *The Journal of Japanese Studies* 33(1): 93–123.

IISD (2014) *Climate Summit 2014: Catalyzing Action.* Accessed on 28 October 2015 from www.iisd.ca/climate/cs/2014/

Kaku, Shishi (2014) 'Chūgoku wa Genshiryoku Shinkōkoku' in Nihon Genshiryoku Sangyō Shimbun, vol. 2716. p. 4. Published 24 April 2014 at www.jaif.or.jp/news_db/data/2014/0424-04-02.html

Kasperson, Roger E., Ortwin Renn, Paul Slovic, Halina S. Brown, Jacque Emel, Robert Goble, Jeanne X. Kasperson, and Samuel Ratick (1988) 'The social amplification of risk: A conceptual framework'. *Risk Analysis* 8(2): 177–187.

Koller, Veronika (2014) 'Cognitive linguistics and ideology', in Jeannette Littlemore and John R. Taylor (eds) *The Bloomsbury Companion to Cognitive Linguistics.* London: Bloomsbury, pp. 234–252.

Lester, Libby and Simon Cottle (2009) 'Visualizing climate change: Television news and ecological citizenship'. *International Journal of Communication* 3: 920–936.

Lester, Libby and Brett Hutchins (2014) 'The power of the unseen: Environmental conflict, the media and invisibility'. *Media, Culture & Society* 34(7): 847–863.

McCombs, Maxwell (1997) 'Building consensus: The news media's agenda-setting role'. *Political Communication* 14: 433–443.

Ministry of Environment, Japan (2012a) *Kikōhendō Tekiō Shimpojium, Kikōhendō Tekiō-shakai e: Chiiki kara no Henkaku* [Towards a climate change adaptable society: Transformation begins at the local communities]. Press Release issued online on 16 October 2012. Accessed on 14 July 2016 from www.env.go.jp/press/15828.html

Ministry of Environment, Japan (2012b) *Heisei 25-nendo Kankyōshō Jūten-sesaku* [Ministry of Environment Priority Measures for 2013]. Accessed on 19 July 2016 from www.env.go.jp/guide/budget/

Ministry of Environment, Japan (2013a) *AR5 Dai-ichi sagōbyōbukai no hōkoku: Kikōhendō 2013, Shizen-kagakuteki konkyo.* Issued online 27 September 2013. Accessed on 27 September 2013 from www.env.go.jp/earth/ipcc/5th/#WG1

Ministry of Environment, Japan (2013b) *Efforts of the Ministry of Environment, Japan, towards Adaptation to Climate Change* [Kikō-hendō no tekiō ni muketa

Kankyō-shō no torikumi ni tsuite]. Presented at the Kikōhendō Symposium on 26 November 2013.

Ministry of Environment, Japan (2013c) *Kikōhendō no mijikana eikyō to tekiōsaku wo kangaeru: "IPCC Dai-38-kai sōkai in Yokohama ni mukete" no kaisai ni tsuite (oshirase)*. Issued online on 7 October 2013. Accessed on 17 July 2016 from www. env.go.jp/press/17219.html

Ministry of Environment, Japan (2014a) *Ekkyō Taikiosen/Sanseiu Chōki-monitaringu Hōkokusho (Years of Heisei 20–24) no Kōhyō ni tsuite (Oshirase)*. Issued online on 27 March, 2014. Accessed on 14 July 2016 from www.env.go.jp/press/press. php?serial=17946

Ministry of Environment, Japan (2014b) *Kikōhendō ni kansuru Seifukan-paneru (IPCC) Dai-5ji Hyōka hōkokusho (AR5) nitsuite*. Issued online 2 November 2014. Accessed on 28 October 2015 from www.env.go.jp/earth/ipcc/5th/

Ministry of Environment, Japan (2014c) *Chiiki-kankyō Symposium "Kikōhendōno kagaku to watashitachi no mirai: IPCC Sagyōbukai kyōdō-gichō wo mukaete" no kaisai ni tsuite (Oshirase)*. Issued online on 25 December 2014. Accessed on 14 July 2016 from www.env.go.jp/press/100154.html

Ministry of Environment, Korea (2014) *Photo News: The 16th Tripartite Environment Ministers Meeting among Korea, Japan and China (TEMM 16)*. Issued on 30 April 2014. Accessed on 14 July 2016 from http://eng.me.go.kr/eng/web/board/ read.do?menuId=22&boardMasterId=523&boardId=354752&boardCategoryId=

Ministry of Foreign Affairs (2013) *'Seme no Chikyū Ondanka Gaikō-senryaku' no Sakutei*, Accessed on 28 October 2015 from www.mofa.go.jp/mofaj/press/ release/press4_000286.html

Ministry of Foreign Affairs (2014) *Statement by Prime Minister Shinzo Abe at the Plenary Session of the UN Climate Summit 2014*. Accessed on 28 October 2015 from www.mofa.go.jp/mofaj/ic/ch/page3_000918.html

Ministry of Foreign Affairs, Ministry of Economy, Trade and Industry, and Ministry of the Environment of Japan (2013) *ACE: Actions for Cool Earth: Proactive Diplomatic Strategy for Countering Global Warming*. Accesssed 19 July 2016 from www.mofa.go.jp/files/000019537.pdf

Miyazawa, Kensuke and Yamada Junji (2015) 'The growth strategy of Abenomics and fiscal consolidation'. *Journal of the Japanese and International Economies* 37(S1): 82–99.

Nagashima, Miori (2015) 'リスクの不確実性と不確実性のリスク～国際的科学評価というサブ政治', in Miori Nagashima, Glenn D. Hook and Piers R. Williamson (eds) *kakusan-suru risuku-no seijisei: soto-naru zashi; uchi-naru zashi* [*The Politics of the Spread of Risk: Internal and External Observations*] 拡散するリスクの政治性. Nara: Kizasu shobō, pp. 37–76.

Nakano, Kaori (2013) PM2.5 をめぐる問題の経緯と今後の課題 立法と調査 2013.10 No.345 (参議院事務局企画調整室編集).

The Nihon Keizai Shimbun (3 February 2016) 'PM2.5硫酸イオン主成分', morning edition.

Pachauri, Rajendra K. and Andy Reisinger (eds) (2007) *Contribution of Working Groups I, II and III to the Fourth Assessment Report of the Intergovernmental Panel on Climate Change*. Geneva, Switzerland: IPCC.

Penza, Michele, Domenico Suriano, Maria Gabriella Villani, Laurent Spinelle and Michel Gerboles (2014) 'Towards air quality indices in smart cities by calibrated low-cost sensors applied to networks'. *IEEE Sensors 2014 Proceedings*. Conference

2–5 November 2014. Accessed on 6 October from http://ieeexplore.ieee.org/document/6985429/

Renn, Ortwin (2011) 'The social amplification/attenuation of risk framework: Application to climate change'. *WIREs Climate Change* 2: 154–169. doi:10.1002/wcc.99

Sørensen, Mette, Herman Autrup, Ole Hertel, Håkan Wallin, Lisbeth E. Knudsen, and Steffen Loft (2003) 'Personal exposure to PM2.5 and biomarkers of DNA damage'. *Cancer Epidemiology, Biomarkers & Prevention* 12: 191–196.

Stavinoha, Agnieszka, and Simona Balbi (2012) 'The use of Network Analysis tools for dimensionality reduction in text mining'. Conference paper presented at *Text Mining Symposium on Learning and Data Science, 8 May 2012, Florence, Italy*. Accessed on 9 July 2016 from www.researchgate.net/profile/Simona_Balbi/publication/273770776_SLDS_Firenze_8_maggio_2012/links/550c2c690cf20637993a0fb5.pdf

Straif, Kurt, Aaron Cohen, and Jonathan Samet (eds) (2013) *Air Pollution and Cancer*. IARC Scientific Publication 16.

United Nations Framework Convention on Climate Change (2015) Accessed on 10 July 2016 from http://newsroom.unfccc.int/unfccc-newsroom/finale-cop21/

Valdés, Juan Gabriel (1995) *Pinochet's Economists: The Chicago School in Chile*. Cambridge, UK: Cambridge University Press.

van der Does Ishikawa, Luli (2013) *A Sociolinguistic Analysis of Japanese Children's Official Songbooks, 1881–1945: Nurturing an Imperial Ideology through the Manipulation of Language*. PhD thesis, University of Sheffield.

van der Does-Ishikawa, Luli (2015) 'Contested memories of the Kamikaze and the self-representations of Tokkō-tai youth in their missives home'. *Japan Forum*, Special Issue: Excavating the Power of Memory 27(3): 345–379.

van der Does-Ishikawa, Luli (2016) 'Cross-border atmospheric pollution and media power'. *Presentation at the TLLP Seminar, Hokkaido University, 28 June 2016*.

van Eck, Nees Jan, Ludo Waltman, Rommert Dekker, and Jan van den Berg (2010) 'A comparison of two techniques for bibliometric mapping: multidimensional scaling and VOS'. *Journal of the American Society for Information Science and Technology* 61(12): 2405–2416. doi:10.1002/asi.21421

Vineis, Paolo and Kirsti Husgafvel-Pursiainen (2005) 'Air pollution and cancer: Biomarker studies in human populations'. *Carcinogenesis* 26(11): 1846–1855.

Weblio (2015) Accessed on 31 January from www.weblio.jp/info/service/thesaurus.jsp#weblio_thesaurus

Wodak, Ruth and Michael Meyer (2009) 'Critical discourse analysis: History, agenda, theory and methodology', in Ruth Wodak and Michael Meyer (eds) *Methods of Critical Discourse Analysis* (2nd edition). London: Sage, 1–33.

World Health Organization (2013) *Health Effects of Particulate Matter: Policy Implications for Countries in Eastern Europe, Caucasus and Central Asia*. WHO Regional Office for Europe, Denmark. Accessed on 27 July 2016 from www.euro.who.int/__data/assets/pdf_file/0006/189051/Health-effects-of-particulate-matter-final-Eng.pdf

World Health Organization (2014) *Air Quality Deteriorating in Many of the World's Cities*. Press release issued on 7 May 2014. Accessed on 27 July 2016 from www.who.int/mediacentre/news/releases/2014/air-quality/en/

Yelland, Phillip M. (2010) 'An introduction to correspondence analysis'. *The Mathematica Journal* 12: 1–23. http://dx.doi.org/doi:10.3888/tmj

6 Visions of a super smart society

Risk management and responsibility as adaptation

The Japanese state is embarking on an economic revitalisation strategy centred on 'unleashing the power of the private sector to the fullest extent', 'creating new frontiers', 'fostering human resources' and 'redistributing the fruits of growth to people's lives' (Kantei 14 June 2013). It promotes science, technology and innovation, with ambitions of 'becoming the world's leading IT society', a 'business hub' to compete on a global playing field, 'improving infrastructure' 'realizing clean and economical energy' and extending the nation's 'healthy life expectancy' (Kantei 14 June 2013). Though broad in scope, within it lies the state's way of responding to pollution, and more broadly climate change, a problem whose scope is much broader in terms of impacts, and whose impacts are poised to intensify (e.g., Meinshausen *et al.* 2009; Chung *et al.* 2014; Rignot *et al.* 2014).

This chapter argues that amidst the responses adopted by states and other actors to climate change, the term 'smart' (*sumāto*) in Japanese has been applied to many projects as a buzzword (Cornwall and Brock 2005; Cornwall 2007; Cornwall and Eade 2010) for development policy, signifying this technological improvement to systems of resource management, which has the potential to fundamentally alter society. For climate change, transition to a low-carbon society 'involve[s] the consideration of uncertainties around future technological and social changes', where an understanding of the relevant dynamics of actors/ institutions facilitates a better understanding of uncertainties and therefore strategic and policy effectiveness regarding such change (Hughes *et al.* 2013). However, while 'smart' solutions to climate change offer exciting avenues of development, the wide application of 'smart' as a buzzword can potentially bring with it changes in the relationship between the state and citizen, changes in citizen's private and professional lives, as well as impact on the climate, *per se.*

Of course, this does not mean that we in any way aim to downplay the necessity of adopting effective responses nor the potential of solutions to the environmental challenge posed by climate change. Rather, *given the necessity of a solution*, this two-part study explores the roles of different actors within the holistic vision of a smart society. It employs a similar theoretical approach to that adopted in Chapter 5 by focusing on the role of actors, employing the state, market and society as a heuristic device (Hook 2012; Mason 2012, 2014;

Hook *et al.* 2015; Nagashima *et al.* 2015) to examine the institutional diffusion (Strang and Meyer 1993; Wakuta 2015) of 'smart' initiatives. Here, it is claimed that alongside agenda setting (Wakuta 2015), the use of buzzwords may be considered a communicative tool used to disseminate a given practice or innovation to a wide range of actors.

The diffusion of 'smart initiatives' is essentially related to 'theorisation', which refers to a semiotic process concerning 'the self-conscious development and specification of abstract categories and the formulation of patterned relationships such as chains of cause and effect' (Strang and Meyer 1993: 492). It denotes general models for understanding a given practice which allow for its diffusion. Diffusion becomes more rapid and extensive through cultural linkages that derive from 'the cultural understanding that social entities belong to a common social category [which] constructs a tie between them', such as a 'society' or 'nation', particularly when the social category is 'informed by theories at higher levels of complexity and abstraction' (Strang and Meyer 1993: 490–493; Wakuta 2015: 5). Further, culturally legitimated theorists are important for the universal diffusion of a given practice (Strang and Meyer 1993: 494; Wakuta 2015: 6). This can be illustrated in Japan (and elsewhere) by the existence of epistemic communities (Haas 1992; Jagers and Stripple 2003) derived from 'industry–academia–government collaborations' (*sangakukan renkei*). Thus, as visions for a new 'society' are constructed and negotiated by groups such as epistemic communities, the use of 'smart' as a buzzword by internal and external actors functions as a mechanism of abstraction from specific, concrete policies, making the concept more general and fuzzy (Vanolo 2014) while at the same time more accepted[1] (e.g. Cornwall and Brock 2005).

Rather than follow the well-trodden path of assessing technological innovations to tackle climate change, this chapter seeks to address their 'political qualities'. It adopts this focus as those innovations designed to overcome the problems facing Japan (and the world more generally) such as climate change potentially constitute a 'means of establishing patterns of power and authority in a given setting' (Winner 1980: 134). In other words, by focusing on 'smart' innovations in Japan, we seek to explore how the introduction of new innovations lends itself to new practices in Japanese society in order to overcome the challenges that face the nation, and how new practices contribute to the changing relationship between various stakeholders in Japan, using the triumvirate of state–market–society actors as a heuristic device.

This chapter focuses on Japan in particular owing to the Cabinet Office's declaration of making Japan a 'super smart society' ahead of the 2020 Tokyo Olympics. It constitutes a vision of society of prioritised resource efficiency and is defined as

> a society where all people can live comfortably and actively, overcoming various differences in terms of language, region, sex, age, receiving high-quality services, and is able to respond attentively to the various needs of society,

supplying only when necessary the necessary goods and services to the necessary people at the necessary time.

(Cabinet Office Government of Japan 22 January 2016: 11)[2]

Further, its scope is considerable with a 'service platform' consisting of eleven, wide-ranging domains:

1   Optimisation of energy value chain
2   New manufacturing system
3   Regional comprehensive care
4   Infrastructure maintenance management and updating
5   Resilient society regarding natural disasters
6   A hospitality system
7   A global environment information platform
8   An integrated material development system
9   Smart production system
10 Smart food chain system
11 Intelligent (*kōdo*) transportation system

(Ishida 27 January 2016)

The 'super smart society' refers to the state's attempts to pioneer socio-economic transformations at the outset of what is called 'the fourth Industrial Revolution' now taking place. This change is due to information communication technology (ICT) development and information sharing, integrating sectors, services and resources, resulting in an increasingly globalised environment (Cabinet Office Government of Japan 19 March 2015). It relates to the 'smart Japan ICT strategy' (*sumāto japan ICT senryaku*), and focuses on the 'Internet of Things' (IoT), robotics/artificial intelligence (AI) and big-data analysis to create an integrated system (Cabinet Office Government of Japan 19 March 2015). Here, the use of 'smart' extends to a wide range of sectors with projects/initiatives such as 'smart service', 'smart agriculture', 'smart regional care', 'smart resilience', 'smart mobility', 'smart infrastructure', 'smart grid' and 'smart manufacturing' (Cabinet Office Government of Japan 19 March 2015).

At the same time, this chapter also seeks to introduce 'smart' innovations, particularly 'smart cities', as a global phenomenon not limited to Japan. For this reason, the next section focuses on 'smart cities' to introduce the topic by discussing concrete plans across the three geographical foci of this book, Australia, China and Japan. Following this is a section on 'Japan's visions of a smart future', which narrows the focus to Japan in order to provide an explorative analysis of 'smart' discourses by examining the perceived role of actors across a broad range of smart initiatives. The aim is to demonstrate how 'smart' as a buzzword for a highly abstract concept operates to engender subtle changes in the relationship between actors while the state, market and individuals endeavour to overcome the challenges facing the nation – particularly climate change – through the optimisation of resource use. The final section concludes the chapter by arguing that the use of ameliorative language associated with 'smart' initiatives can serve to encourage citizens to take on additional roles and responsibilities with certain initiatives encouraging, for example, behavioural adaptation. Further, it is argued that this and the flexible application of 'smart' potentially affords greater structural flexibility for the state to respond to the challenges ahead, but also has the potential for the off-loading of state risk on to the market or society amidst the changing relationship between the state and citizen.

Smart cities and climate change: Australia, China and Japan

As stated above, this chapter focuses on 'smart' initiatives amidst responses to pollution, and more broadly climate change, and other threats to national sustainability, as well as introducing such initiatives as a growing global movement. To this end, this section offers a brief overview of investment into 'smart cities' in Australia, China, and Japan, before providing a detailed examination of the Japanese case.

While many argue that smart cities have the potential to respond effectively to the deleterious effects of climate change, they do not relate solely to this issue but also harbour the potential to greatly influence the whole gamut of the citizen's everyday life (Vanolo 2014). For instance, the website smart-cities.eu offers a means to evaluate the performance of European smart cities based on the European Smart City Model, which was developed 'in cooperation with different partners and in the run of distinct projects financed by private and public stakeholders and actors' (Technische Universität Wien 2015a). Measurements include 'smart economy', 'smart people', 'smart governance', 'smart mobility', 'smart environment' and 'smart living' (Giffinger *et al.* 2007; Technische Universität Wien 2015b). Visions of a 'smart' city/town/society are part of a growing global movement (Dirks *et al.* 2010), albeit still ill-defined (Giffinger *et al.* 2007; Vanolo 2014). In Australia, China and Japan, they constitute a means to respond to climate change, amidst other national threats, as well as a model of economic growth. A number of examples of initiatives in Australia, China and Japan related to smart cities are provided below to explain their potential scope.

Australia

A consortium of consultancy firms was commissioned to assess smart grid technologies in Australia, promoting the potential economic benefits of smart grids (Arup 2014: 1–2). Among these, Arup notes that the government's A$100- million 'Smart Grid, Smart City' initiative stems from the transition from 'highly positive growth' to 'negative growth', behoving the state to change its national energy sector (Arup 2014: 4).

The programme amounted to A$490 million from all contributors, and it was predicted that the use of smart grid technology could result in a net economic benefit of up to A$28 billion over the following two decades (Arup 2014: 5).

To this end, the report recommends that alongside smart grids, 'cost reflective electricity pricing', 'consumer behaviour change with respect to electricity consumption', including 'consumer education', and 'energy market reform' are necessary (Arup 2014: 5/86–87). Amidst such change, ICT is frequently incorporated as a core component of smart cities because it facilitates more efficient management of the environment, infrastructure and resources (Bajracharya *et al.* 2014: 119). To assist in construction, three city councils in Australia were awarded an IBM grant as part of its Smarter Planet initiative which promotes the application of ICT to increase the efficacy and efficiency of city energy systems and interconnectivity of 'smart' devices (Bajracharya *et al.* 2014: 120).

Here, 'Internet of Things', denoted as pervasive internet and industrial internet with 'the interconnection of objects to internet infrastructure through embedded computing devices [. . .]' (World Bank 2016: 328), may be applicable to 'every single sector of society and the economy', creating 'new businesses' and 'new industries' and combatting climate change (PR Newswire 19 March 2015; World Bank 2016: 328–329). Further, it is predicted that by 2025 technological innovations will lead to 'augmented reality' enhancing perceptions of 'real-world displays' through 'portable/wearable/implantable technologies'; disrupting twentieth-century business models, affecting education, among other sectors; and the tagging, databasing, and analytical mapping of physical and social realms (Australian Communications and Media Authority 2015: 5). They also predict 'a global, immersive, invisible, ambient networked computing environment built through the continued proliferation of smart sensors, cameras, software, databases, and massive data centers in a world-spanning information fabric' (Australian Communications and Media Authority 2015: 5).

Direct economic contribution of communication services alone in Australia amounted to A\$21 billion in 2011 and is expected to double by 2031 (Australian Communications and Media Authority 2015: 6). To pioneer this, Cisco Systems established an industry and research collaboration centre with Curtin University and Woodside Energy. The centre's aim is to 'create a state-of-the-art connected community in Western Australia' which is 'focused on leveraging cloud, analytics, cyber security and (IoE)[3] network platforms' with the view of later interconnecting with the east coasts (Nicholls *et al.* 30 July 2015).

China

Chinese smart city (*zhineng chengshi* or *zhihui chengshi*) projects are also on the rise and represent a pathway to scale-up green energies which theoretically curtail the need to import finite energy resources whose combustion emits harmful pollutants. As touched on in Chapter 4, this is no easy task as coal amounts to over 70 per cent of China's primary energy consumption (Zhang *et al.* 2015: 1742). However, the use of renewables may strengthen China's energy security where there is less reliance on fossil fuel imports. To this end, the (smart) Asian Super Grid – a proposal to link the electricity grids of China, Japan, South Korea, Mongolia and potentially Russia and, more speculatively, other states in

Southeast Asia, with the application of ICT to moderate fluctuating levels of supply and demand – may be considered a means to scale-up the use of renewable energies (Mathews 2012). Additionally, China is building the world's largest renewable energy system and spent US$4.3 billion on 'smart grid' technology, constituting almost 30 per cent of the world total (Mathews and Tan 2014).

These also constitute an economic growth model. For instance, China's Ministry of Industry and Information Technology (MIIT) places 'smart restructuring' (*zhineng zhuanxing*), alongside 'green development' (*lüse fazhan*) as a key component of the '2025 China Manufacturing' (*2025 Zhongguo zhizao*) initiative (Ministry of Industry and Information Technology, China 19 May 2015). MIIT claims that technological innovations profoundly affect 'the competitive landscape of world powers' as 'the rise and fall of manufacturing confirms the rise and fall of world powers' (Ministry of Industry and Information Technology, China 19 May 2015). Moreover, claiming to have learned the lessons of history twice before, the initiative responds to a new industrial revolution requiring the mobilisation of all sectors of society (Ministry of Industry and Information Technology, China 19 May 2015). With this, China Unicom has signed up for the construction of smart projects across 150 cities nationwide. China Telecom and twenty-eight provincial governments have signed for the construction of 179 smart cities. China Mobile is reportedly seeking to construct Chinese 'wireless cities' (*wuxian chengshi*) (Smart City China 20 January 2016). These smart city projects entail large-scale investments in social infrastructure, with the website smartcitychina.cn providing an overview of smart city functions:

- Harmonisation of public services
- Integrated financial system in each city
- Unification of environmental protection
- Security monitoring
- Integrated telecommunications system in each city
- Integrated public transportation

(Smart City China 20 January 2016)

As an illustration of the extent of Chinese initiatives, on 29 March 2016 the State Grid Corporation of China (SGCC) co-hosted the 2016 International Conference for Global Energy Interconnection in Beijing, which was sponsored by the China Electricity Council, China Society for Electrical Engineering and the China Machinery Industry Federation, among others (State Grid Corporation of China 30 March 2016). The event was the first international conference dedicated to global energy interconnection, where SGCC chairman Liu Zhenya stated that a 'world grid' could be running by 2050 (Spegele 30 March 2016; State Grid Corporation of China 30 March 2016). Although this may appear somewhat anecdotal, in 2015 the National Energy Administration of China (NEA) issued the 'Distributed Network Construction and Reform Action Plan (2015–2020)' (*peidianwang jianshe gaizao xingdong jihua*) with investment in 2015–2020 considered to be no less than 2 trillion Chinese yuan (CNY), and is set to rapidly accelerate smart grid development, with preliminary estimates suggesting China's peak electricity load to will reach about 1.4 times that of 2014 by 2020 (Chen 7 September 2015).

Japan

Smart city rollouts have a unique context in Japan where renewable energies were to become more of a priority following the Tōhoku Earthquake and Fukushima Nuclear Disaster (3.11) (DeWit 2011). For context, '3.11' denotes the tragedy brought about when an earthquake measuring 9.0 on the Richter scale (the Great East Japan earthquake) occurred 200 kilometres off the San-riku Coast on 11 March 2011. The subsequent tsunamis killed approximately 20,000 people, displaced approximately 390,000 people and caused the largest nuclear meltdown in world history, the effects of which are in some cases still being felt (Ikeda 2013: 15; Nagasaka 2013: 7; Sekizawa 2013) and yet to be fully understood (Williamson 2014).

Following 3.11, Masayoshi Son, chairman of the telecommunications giant SoftBank Corp, established the 'Japan Renewable Energy Foundation'[4] (Nozawa 21 April 2011). Along with renewable energies, Son has also pro-moted 'feed-in tariffs' aimed at utilising locally produced power alongside other sources, conducive to renewable energies and smart grids, designed to enable energy provision from distributed energy supplies (DeWit 2011, 2014b). Alongside popular demonstrations against the restart of nuclear reactors, the aftermath of 3.11 bore witness to a nationwide campaign to reduce electricity consumption by 15 per cent, and attained a remarkable 20 per cent decrease overall[5] (Kingston 2012).

The transition to alternative energy sources links to smart cities projects, many of which have transitioned from experimental trial to deployment in Japan, and involves a broad range of state and market actors (Kinjō and Rure 2014), with public and private initiatives (DeWit 2014a). These projects reportedly centre provision around citizens' needs through the business model of supply and demand (Kinjō and Rure 2014: 9). They may be used to create 'low-carbon cities' (Ministry of the Environment, Japan 12 April 2013: 49), and with claims of abundant renewable energy resources in Japan (Lovins 8 July 2014) the drive for renewable energies is also predicated on laying the groundwork for a new economic model (Kinjō and Rure 2014: 7). The key question is thus: what exactly is 'smart' about smart cities? Here, the Ernst & Young Institute provides the following definition:

1   'Improving the convenience and comfort of citizens' lifestyles';
2   'Pursuing sustainability (through autonomy, environmental consideration, economic growth and safety)';
3   'Making services, and enterprise [including infrastructure] more efficient in terms of time and resource' (i.e. 'people, goods, money').

(Kinjō and Rure 2014: 3)

Crucially, smart cities ostensibly aim to construct a 'win-win-win relationship between the three stakeholders of state, business and society' (Kinjō and Rure 2014: 6). With this, ministries have sought budgetary increases for renewable energy/energy conservation projects alongside the 'Internet of Things' initiatives (DeWit 2015; see also Ishida 1 September 2015; Kankyō Bijinesu Onrain 2 September 2015; Ministry of Economy, Trade and Industry, Japan August 2015; Ministry of the Environment, Japan August 2015; Ministry of Internal Affairs and Communications, Japan August 2015).

However, this does not necessarily entail a complete transition to renewable energies as a means to combat climate change. For instance, the Abe cabinet's 'hydrogen economy', marketed as part of a 'low-carbon society' ahead of the Tokyo 2020 Olympics, may be heavily reliant on Australian low-grade carbon because Japan has no domestic source of hydrogen gas (Hanley 18 September 2015 in DeWit 2015). Further, smart cities do not necessarily preclude nuclear power (Kingston 2012), with small model reactors potentially affording the generation of electricity in remote areas with limited to no access to the power grid (Nuclear Energy Institute 2015), and so – from a cursory viewpoint at least – appear more compatible with smart grids, and a means for revenue.

Overall, smart initiatives have centred on the need to 'do more with less' by increasing the 'three Es' of 'efficiency' (*kōritusei*), 'effectiveness' (*yūkōsei*) and 'economy' or financial stability (*keizaisei*) of the existing system by expanding ICT into the national infrastructure to integrate and optimise facilities. The drive for the 'three Es' has been a prominent theme in the public administration of many states, including Japan. For example, the 'smart shrink' (*kashikoi gyōshū* or *sumāto shurinku*) model for society aims to streamline services 'smartly' as the labour force declines where, strategic investment of a town may be accompanied by the integration, abolition or sale of fungible public facilities, though should only take place with public consent (Hayashi 2011; Kasahara and Iwamoto 2015). The need to 'do more with less' also emerged during the 'New Public Management' (NPM) wave in public administration that laid emphasis on adopting (quasi-) market mechanisms to public services whether through privatisation, deregulation or externalisation through public–private partnerships (PPPs) or private finance initiatives (PFIs) (Andrews and van de Walle 2012; Kudo 2004: 155–156; Yamamoto 2003). Accompanying this has been the Japanese state's transfer of risk to the individual through policies aimed at fostering autonomy and individual initiatives in order to support the emergence of 'self-responsibility' as a norm (Hook and Takeda 2007; Takeda 2005). With this, the following analysis seeks to examine the role of the 'self' – that is, the individual – in 'smart' initiatives amidst these changes.

Japan's visions of a smart future

As detailed below, 'smart' is used as a buzzword relating to development policy (Cornwall and Brock 2005; Cornwall 2007; Cornwall and Eade 2010).

Development buzzwords are considered to be more than jargon and serve, like the notion of development itself, to convey the sense of a better tomorrow, leaving 'much of what is actually *done* in its name unquestioned' (Cornwall 2010: 2, emphasis in original). Buzzwords are often vague, euphemistic, flexible in application and normative resonance, sound intellectual, scientific, best left to the 'experts', while also defining what's in vogue (Cornwall 2010: 2–3). Crucially, they do not simply cloud meaning but also combine performative qualities with 'an absence of real definition and a strong belief in what the notion is supposed to bring about' (Cornwall 2010: 3; Rist 2010: 20).

It is argued here that 'smart' can be considered a development buzzword according to the above definition in its conventional application. Therefore, in line with the theoretical approach, this section aims to explore what roles it entails for the state, market and society. In order to do so, the analysis is carried out through the following methodology:

1   The definition of 'smart' is examined to assess its overall meaning.
2   Its application in the bicameral National Diet is assessed using data gathered from the search engine of Diet proceedings[6] (National Diet n.d.). This is carried out in three steps:

    a   An overall view of the term's frequency of use from 1 January 1951 to 31 December 2015 is provided.
    b   An analysis of the referent of 'smart' is carried out based on a.
    c   The most prevalent terms are defined and their emergence examined to give an overview of 'smart' concepts discussed in Japan's legislature.

3   Texts from outside the Diet relating to the most prevalent 'smart initiatives' are discussed within the context of initiatives aimed at encouraging change in individual behaviour such as those relating to consumption habits. The analysis of the findings is split into two sections. Each section introduces the most prevalent initiatives discussed in the Diet, while section two also incorporates an external concept, healthcare, to demonstrate the adaptability of 'smart' to an array of contexts which signify changes to individual lifestyles.

This section explores each stage in order; that is, stage one in the section 'Defining "smart"', stage two in 'Applications of "smart" in the Diet' and stage three in the section '"Smart life" and the individual: textual analysis'.

Defining 'smart'

The adjective 'smart' has four meanings:

1   A good appearance in terms of one's figure or the shape of something looking slender.
2   Quick and refined in terms of behaviour, etc.

3   Tasteful in terms of dress-sense.
4   High tech in terms of electronic devices.

(Goo Jisho n.d.)

Similarly, it can be defined as 'one's attire or movements appearing refined' or 'slender in terms of figure' (Sanseido n.d.). Weblio provides a number of synonyms for 'smart', categorising them into semantic groups denoting stylishness, high intellect, or a slender figure (Weblio n.d.). The term 'smart' conventionally denotes four potentially overlapping conceptual domains, namely sleekness, slenderness in appearance, stylishness, intelligence, and technological sophistication. In short, it is a positively loaded judgement which describes an above-average condition of a given object in terms of physical appearance, behaviour, design or intellect.

Applications of 'smart' in the Diet

Figure 6.1 shows the number of proceedings in the Diet containing the term 'smart' from 1951 to 2015 grouped into five-year periods. For example, in 1950 there were four proceedings from the Lower House that contained the word 'smart': 'the Committee on Education' (*monbu iinkai*), 'Special Committee Related to the Repatriation of Overseas Countrymen' (*kaigai dōhō hikiage-ni kansuru tokubetsu iinkai*), 'Telecommunications Commission' (*denki tsūshin iinkai kōchōkai*), and 'Committee on Rules and Administration' (*giin unei iinkai*). All speeches, interpolations, questions, remarks, and so forth, within one session were taken together so that, irrespective of who said 'smart' or how many times it was said, the year 1950 would be quantified as four because 'smart' was used in four separate sessions.

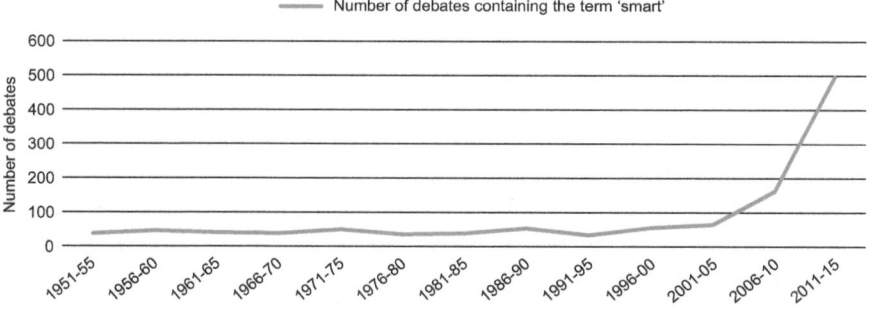

Figure 6.1 Number of debates containing the term 'smart' from 1951 to 2015
Source: Author

'Smart' was used in approximately fifty sessions every five years from 1951 until the turn of the millennium. Following this, sessions containing the term 'smart' increased approximately tenfold. To investigate the application of the term in more detail, every use of 'smart' from 1 January 2000 up to 31 December

2015 was analysed to find its referent. Uses as adverbs or predicative adjectives were excised from the analysis owing to possible ambiguity. There were a number of terms where 'smart' was 'in-fixed' between nouns, such as 'Japan Smart Life Project'. These were included as there is no ambiguity over reference. These were then categorised into conceptually related groups, the full list of which is provided in Table 6.1 in the Appendix below.

In order to better understand this increase in the application of 'smart', its use within the five most frequent categories in every year starting from 2000 up to the end of 2015 is provided in Figure 6.2. This relates not to proceedings *per se*, but rather the number of times the term 'smart' relating to the categories presented below was spoken.

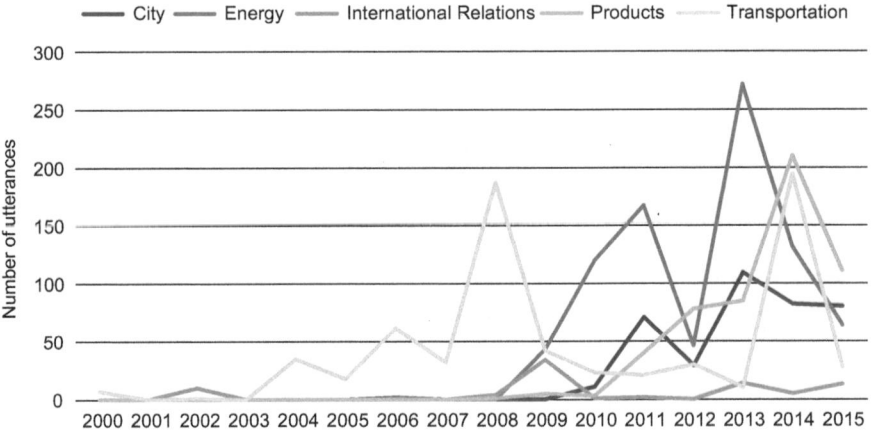

Figure 6.2 Application of 'smart' during individual sessions in the Diet
Source: Author

The use of 'smart' increased from 2004, when its application was related to transportation. From 2004 to the end of 2015, all but two 'smart' references within the Transportation Category referred to facets of the smart interchange relating to Japan's roadways, while in 2015 'smart support stations' in Japanese railway stations were mentioned twice. 'Smart interchange' became a major talking point in the Diet in 2008, while other 'smart' initiatives, such as in the Energy Category, emerged between 2010 and 2011, peaking in 2013. Undulations in the City Category mirror those of the specific Energy Category, signifying their conceptual relatedness (see Appendix 6.1 for full table of categories and references). Products emerged in 2012 and peaked in 2014, alongside the 'smart interchange'.

Although it is not evident in Figure 6.2, 'smart'-related terms within a category diversified. For instance, in 2010 and 2011, when terms within the 'City Category' began to emerge on a consistent basis, 'Smart Community' (mentioned 61 times, i.e., $n = 61$), 'Smart City' ($n = 9$), 'Smart House' ($n = 2$) and

'Okinawa Smart Island Initiative' (n = 1) comprised the terms used. However, in 2015 the category had broadened to include 'City' (n = 20), 'Community' (n = 23), 'Community City' (n = 2), 'Eco Park' (n = 4), 'House' (n = 3), 'Platinum Society' (n = 1), 'Residence' (n = 13), 'Town' (n = 2), 'Town/ Community Planning' (n = 2) and 'Wellness Residence' (n= 5), indicating a broadened and more flexible application of the 'smart' concept.

Despite this, within every major category there were terms that were used far more than others. Here, the two most frequently used term in each category are provided in Figure 6.3. 'Smartphone' is the only term in the Product Category because it is by far the most frequently used term in this category, with 'smart TV' the second most frequent (n = 31). Further, the 'International Relations' category was excised because the frequency of individual terms is too low and inconsistent to discuss the emergence of a key term.[7]

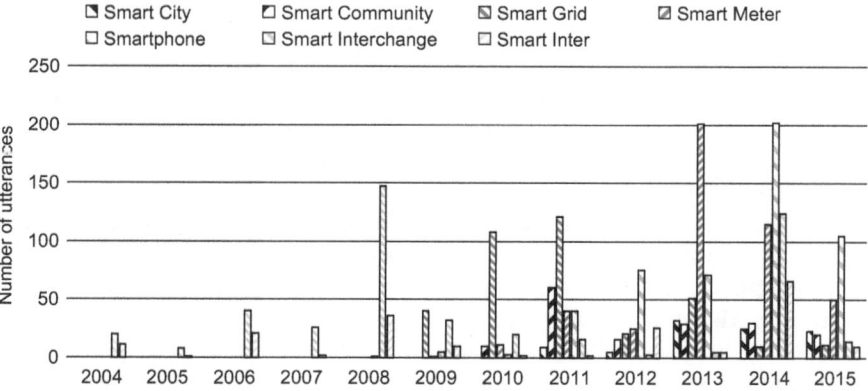

Figure 6.3 Most frequently used 'smart' phrases in the Diet debates (2004–2015)
Source: Author

Here, 'smart interchange' was a major talking point until 2009 when 'smart grids' emerged, dramatically increasing in the subsequent two years which also saw the emergence of 'smart community/city'. Debate shifted over 'smart grids' and onto 'smart meters' in 2013 while 'smartphones' and 'smart interchange/ inter' would increase from 2011 to 2014 also.

The rise of smartphones is likely explained by their emergence in the Japanese market to compete against the highly specialised domestic models. Like a wide range of Japanese ICT products, cellular handsets grew as an industry confined to the domestic market, though highly sophisticated with pioneering features predating Apple's iPhone and smartphones (Kushida 2011). However, as a result, they developed in a separate trajectory within this isolated ecosystem, decoupling Japanese models from global markets (Kushida 2011). Following the launch of smartphone models from 2008, these domestic models were essentially pushed aside by global brands, with the gradual market penetration of Apple's iPhone through three major phone operators alongside Android OS smartphones (Shinohara *et al.* 2013; Kushida 2015: 49).

However, the rise of smartphones is predated by separate categories of 'Interchange', 'Energy' and 'City'. Of these, 'smart interchange' and 'smart meter' were the most commonly used terms. Due to this, a brief explanation of these initiatives and their emergence is provided below.

What is a smart interchange?

A 'smart interchange' denotes an automated tollgate operated by an electronic toll collection (ETC) system (Hirai *et al.* 2005). The Japanese expressway system contains toll roads which were established following World War II to expedite the construction of a nationwide expressway network and were intended to become toll-free following the repayment of construction debts (Mizutani and Uranishi 2006: 1). The smart interchange represents a partial cost-cutting exercise as it lowers labour costs because transactions take place with the automated scanning of an ETC card displayed in the vehicle, forgoing cash-based transfer and the staff requisite to handle cash-transactions. ETC systems have also been extended to service areas (SAs) and parking areas (PAs) (Hirai *et al.* 2005).

Although smart interchanges have ostensible economic benefits, they are considered to provide a means to respond to climate change, albeit less obviously and, potentially, extensively than smart meters, which are discussed in the next section. For instance, the ETC system may contribute to combatting climate change as it reduces tollgate congestion and improves fuel efficiency, with the Ministry of Land, Infrastructure, Transport and Tourism (MLIT) estimating it has reduced nationwide annual CO_2 emissions by approximately 220,000 tonnes (Highway Industry Development Organization August 2014: 596–597). Also, it relates to a broader project of implementing ICT to roadway infrastructures and technologies. For example, the construction of an 'intelligent transport system' (ITS) (*kōdo dōro kōtsū shisutemu*) affords a pathway towards the minimisation of harmful pollutants emitted by vehicles, with breakthroughs in battery storage technology (e.g. Hitt 20 April 2016; Satell 1 April 2016), and the further optimisation of flow dynamics through traffic engineering and fuel consumption (Highway Industry Development Organization 2011b). Additionally, minimising road accidents comprises a major component of this national drive towards a more integrated transportation system (Highway Industry Development Organization 2011b). Here, the emergence of energy efficient/electronic, semi-autonomous cars, equipped with cameras and sensors to assist in safe driving, also termed 'smart cars' (*sumāto kā*) (*Nikkei Shimbun* 3 April 2015), may be considered a market-oriented method of disseminating new technologies to create a system where there are no more accidents,[8] while, simultaneously, affording greater fuel efficiency (Highway Industry Development Organization 2011b: 3, August 2014).

Smart interchange investment is heavily reliant on investment through PPPs and PFIs to improve the three Es of efficiency, efficacy and economic sustainability of roadway management and infrastructure (Highway Industry Development Organization August 2014: 654), extending its scope beyond simply responding to climate change but also potentially engendering new means of administration. For instance, in 2004 MLIT conducted trials and in 2006 over half of the smart interchanges trialled were made permanent (Hirai *et al.* 2005; Asahi Shimbun Company 22 September 2006). In 2005, the Japan Highway Public Corporation (JHPC) was privatised and disaggregated into three expressway companies – East, Central, and West Nippon Expressway Company Limited (NEXCO), respectively – while the Metropolitan Expressway Company Limited (MECL), the Hanshin Expressway Company and the Honshū-Shikoku Bridge Expressway Company were also formed from formerly public organisations (Mizutani and Uranishi 2006). One of the objectives here is to repay interest-bearing debts which amount to more than 40 trillion yen within 45 years, carried out by the Japan Expressway Holding and Debt Repayment Agency (JEHDRA) (Oi 10 September 2012). Thus, while tollgates are used to mitigate congestion and environmental degradation (Hirai *et al.* 2005), in response to climate change, they are, as before, used to repay debt, with the roads said to become toll-free following repayment (*Japan Times* 2 October 2015).

Here, smart interchanges play an important role, with JHPC making use of private consultancy expertise in 2004 to inquire into the expansion and increase of multifunctionality of PA and SA projects (Akiyama 15 July 2006). The resultant report stressed the necessity of capturing expressway revenue opportunities and market expansion into areas external to expressways, while balancing public commitments (*kōkyōsei*) with 'customer-orientation' (*kokyaku shikō*) (Akiyama 15 July 2006) – a logic of public administration related to NPM. Owing to the view that the rise of smart interchanges will increase SA and PA usage (Akiyama 15 July 2006), the newly privatised companies are incentivised to develop, expand and diversify operations using an ETC system, which will increase the scale of reduction of CO_2 emissions (Highway Industry Development Organization August 2014).

Further, the revision in 2008 of the Law Concerning Special Measures for National Financing of Road Construction will lead to additional smart interchanges with toll rate discounts until 2014 (Japan Expressway Holding and Debt Repayment Agency 2014), which aims to encourage users to use the ETC system; both years representing two peaks in parliamentary discussion surrounding smart interchanges. In 2014, the Law for the Partial Revision of the Road Act etc. (*dōrohō-nado no ichibu wo kaiseisuru hōritsu*) was enacted, increasing the tolling period an extra fifteen years to 2065 to repay estimated expressway construction and maintenance costs, including smart interchanges (Ministry of Land, Infrastructure, Transport, and Tourism, Japan 2014; *Japan Times* 2 October 2015). It also permitted the potential private use of underpasses as a means of increasing revenue and securing financial resources for the management of expressways (Ministry of Land, Infrastructure, Transport, and Tourism, Japan 2014). Finally, it mandated an interest-free loan scheme whereby JEHDRA uses state-granted subsidies to loan to expressway companies without interest as part of the funds allocated to developing smart interchanges, with assets and liabilities assigned to JEHDRA following construction (Ministry of Land, Infrastructure, Transport, and Tourism, Japan 2014).

As of 31 December 2015, eighty smart interchanges have been opened that are operated by the NEXCO companies (Ministry of Land, Infrastructure, Transport, and Tourism, Japan 31 December 2015a, 31 December 2015b), while ETC systems are used by the other newly privatised companies (Hanshin Expressway Company Limited n.d.; Honshu-Shikoku Bridge Expressway Company Limited 2005; Metropolitan Expressway Company Limited n.d.). These are likely to increase, with targets of approximately 200 smart interchanges to be developed nationwide by 2018 (Highway Industry Development Organization 2011a), further reducing polluting CO_2 emissions.

What is a smart meter?

A 'smart meter' is a computerised replacement of an analogue meter for a 'smart grid' (McDaniel and McLaughlin 2009: 75) and contains a two-way communication system, allowing meters to disconnect users via software, record usage and send readings to operation centres (Chu *et al.* 2013: 369; Cormack 24 February 2016). Although there are smart water meters, most research and investment relates to electricity systems (Darby 2013: 111). Greater visualisation of energy consumption facilitates greater awareness over consumption habits and hence a potential means by which energy consumption can be optimised through behavioural change to respond and adapt to climate change.

The system's architecture affords the collection and analysis in near-real-time of power transmission, distribution and consumption, and the regulation of consumption of home appliances (Go *et al.* 2013). Further, a net-based metering system is considered more conducive to microgenerators and renewable energies (Darby 2013; Gopakumar *et al.* 2014). Here, distributed generation units such as wind or solar farms can export power to the main grid, while smaller-scale generation associated with commercial or domestic loads are typically designed to operate connected to the main grid or in an 'autonomous' islanded mode in case of grid contingencies (Gopakumar *et al.* 2014: 418–419, 423). As a result, there are claims that smart grids provide a more interactive role for users as both consumers and producers who can add distributed generation sources for greater energy efficiency where excess can be exported to the main grid (McDaniel and McLaughlin 2009; Phuangpornpitak and Tia 2013). This redefinition of roles has led to claims that a 'smart system' demands 'smart meters, smart grids, smart clouds, and smart people' (Shaw 27 February 2016), suggesting how combating climate change through 'smart' technology may affect individuals' behaviours and practices.

The rapid rollout of smart meters may be attributed to a seemingly beneficial relationship between actors intersecting the state, market and society. Suppliers can expect to reduce operational overheads (Rogai 2006 in McKenna *et al.* 2012: 807), and smart meters may facilitate greater market penetration of low-carbon technologies, benefitting transmission and distribution network operators (Strbac *et al.* 2010 in McKenna *et al.* 2012: 807). Also, the use of smart meters may enable governments to meet their carbon-reduction targets (Department for Energy and Climate Change 2010 in McKenna *et al.* 2012: 807) and possibly facilitate lower energy bills for individuals who attain greater energy efficiency (Darby 2006; Owen and War 2006; MacDonald 2007 – all in McKenna *et al.* 2012: 807).

However, alongside privacy issues concerning the scope of information and who is able to access it, whether commercial or criminal third parties, or law enforcement agencies, among others (see McKenna *et al.* 2012; Chu *et al.* 2013; Heather 12 February 2015), the rollout of smart meters has, in places, been problematic. In Australia, for example, the government of Victoria mandated the rollout of 2.6 million smart meters which commenced in 2009 where, though similar benefits to those above were projected, two government reviews found that it had not benefited consumers who, as of 2015, were still paying the costs (Doyle *et al.* 2015). Also, it is reported that the closure of the solar bonus scheme where households can sell electricity back to the grid at a fixed rate forced 130,000 households in New South Wales to pay up to A$600 to install smart meters to avoid hugely increased electricity bills (Nicholls 8 March 2016). Additionally, in China, nearly 300 million smart meters (*zhihui-xing dianbiao* or *zhineng dianbiao*) have been installed as of 2015, with smart meters beginning to replace analogue meters from 2009 (National Energy Administration of China 27 July 2015; Transmission and Distribution World 9 July 2015; China Market Research 5 May 2016). However, a lack of uniformity in standards appears to have hindered somewhat the regulation potential of smart meters, with NEA alongside SGCC and China Southern Power Grid setting a series of smart meter standards within the energy sector in 2015 (National Energy Administration of China 27 July 2015).

In Japan, the rollout of smart meters followed the amendment to the 2014 'Energy Conservation Act' (*shō-ene hō*). The law has been instrumental in incentivising the management of the 'three Es' in terms of energy use among strategic market sectors, enabling the state to respond to pollution and wider climate change domestically and also internationally, dovetailing as a model of growth, with the potential of brand development of technological sophistication and efficiency of Japanese products for the international arena. The law was first enacted in 1979, incentivising energy efficiency among certain industries following the oil shocks (Asahi Shimbun Company 4 March 2009). It was amended in 1999 alongside a number of electricity market liberalisation measures, with the state initiating a 'top runner method' to certain industrial sectors (Asahi Shimbun Company 25 February 1998), taking the highest standard of the most energy-efficient product and making it the standard to be met across the entire sector over a given time frame with penalties for those who do not meet it (Schreurs 2015: 140). Another amendment expanded the framework to other sectors from 2009 and 2010, with estimations that it increased the amount of companies subjected to regulation from 10 per cent to 50 per cent (Kankyō Bijinesu Onrain 2012).

This was accompanied by the 2012 Act on Special Measures Concerning Renewable Energy Electric Procurement by Operators of Electric Utilities (*denki jigyōsha-ni yoru saiseikanō enerugī denki-no chōtatsu-ni kansuru tokubetu sochi hō*) which incentivises the use of renewable energy, strengthening international competition and economic revitalisation of regions within Japan (e-Gov n.d.). Although the frequency of the terms 'smart grid' and 'smart meter' were relatively low in 2012, 2011 was a peak year for 'smart grid'. The act obliges operators of electric utilities to purchase electricity generated from renewable energy sources at a fixed price and for a fixed time frame set by the government (Kojima 24 March 2012; EandE Solutions Inc. 1 August 2012). High energy efficiency and the transition to renewable energies embodies the Japanese state's attempt to respond to climate change, and, crucially, a model for economic growth. The law helped develop feed-in tariffs where operators purchased any excess power generated by microgenerators from renewable sources (DeWit 2012; Harlan 4 June 2013). This also influenced the rapid domestic growth of the solar power industry, presaging a sizeable increase in solar power production, particularly among residential producers (DeWit 2012; Shahan 5 April 2012).

Following this, a further amendment to the Energy Conservation Act obliged electricity suppliers to design and publish plans to introduce smart meters along with a price plan (Chubu Electric Power Company Inc 15 April 2014) as another step towards the scaling-up of renewable energy use in response to climate change. As a result, all ten major electric power companies under the Federation of Electric Power Companies in Japan (FEPC) published plans to introduce smart meters (Chubu Electric Power Company Inc 26 November 2013; Kansai Electric Power Company Inc 17 March 2014; Energia April 2014; Hokkaido Electric Power Company Inc 10 April 2014; Kyuden 30 April 2014; Okiden 30 April 2014; Rikuden n.d.; Yonden 25 November 2014; Tohoku Electric Power Corporation Inc 7 January 2015; Tokyo Electric Power Company Inc 2016). This was followed by an amendment to the Electricity Business Act in 2015, mandating the deregulation of the power sector over the next five years. Here, the regional monopolies of all FEPC companies were lifted, so that citizens choose electrical suppliers in a bid to lower prices through market competition.[9] Power generation and transmission sections within these companies is also due to be disaggregated in 2020, in a bid to increase competition between FEPC and non-FEPC companies (Asahi Shimbun Company 18 June 2015; *Japan Times* 5 July 2015; *Ryūkyū Shimpō* 18 January 2016, 29 January 2016). The market-oriented approach aims to respond to climate change by decentralising the electricity industry, encouraging market competition through disaggregation and market liberalisation, facilitating the market penetration of renewable energies and reducing electricity prices, with the government set to repeal regulations on the retail-sale rates to households after 2035 (*Nikkei Shimbun* 17 June 2015).

'Smart life' and the individual: textual analysis

This section examines the use of 'smart' as a buzzword. The analysis thus far has provided an overview of Diet discussions relating to smart projects, where the 'smart interchange' and 'smart meter' were identified as the most prominently used 'smart' terms in the National Diet since 2004.

Here, in line with the theoretical approach, it is assumed that smart projects at a higher level of abstraction and related to broad common social categories, such as the 'super smart society', are considered to allow for a more rapid diffusion of practices. The dissemination of new practices relates to state, market and societal actors where new practices ultimately become naturalised, with discourses surrounding pollution and climate change more widely being no exception to this (e.g. Chapter 5).

The focal issues here relate to a 'smart life', focusing on the role of the individual and how market mechanisms are seen to facilitate social change. This analysis seeks to examine how notions of a 'smart life' – that is, aspects of individual behaviour – relate to responses to climate change that are centred on the 'smart' individual, as opposed to smart initiatives or smart innovations. With this, the analysis first seeks to demonstrate that notions of a 'smart life' are part of a wider range of initiatives not only related to climate change responses, but a host of other issues facing the state also, with 'smart' used as a buzzword under visions of a 'smart society' to facilitate mechanisms of governance aimed at, among other things, exacting social change in the individual. These include notions of 'smart driving' and 'smart working', both of which relate to combating pollution and climate change, and 'smart living', casting a wide-net over the everyday. This is because it aims to demonstrate and link similarities across a broad range of social domains affecting individuals' lives and shared by notions of 'smart living' to initiatives aimed at effectuating behavioural change regarding energy consumption as a clear and direct response to climate change. In doing so, it is argued that the wide application of smart – whether to initiatives, innovations or individuals – potentially generates political space to off-load state risk onto other sectors as a political act, while the state seeks a new avenue of national sustainability amidst climate change and other issues.

To start, the application of 'smart' as a buzzword covers a range of uses, as illustrated when NEXCO East established the 'Road Control Centre' (*dōro kanri sentā*) in Saitama City in 2016 to centralise and manage roadway information across eastern and northern regions of Japan as a means to optimise traffic flow and expressway safety, with aims of establishing a backup centre in Tōhoku, the area most affected by 3.11 (Daily Engineering and Construction News 9 December 2014; Sankei News 28 February 2016). The project relates to a much broader 'smart' initiative propounded by NEXCO East and the 'SMH Promotion Strategy Committee' (*SMH suishin senryaku kaigi*) to establish a 'smart maintenance highway' (SMH) (*sumāto mentenansu haiuei*) – relating to innovations such as the smart interchange – which entails the use of ICT to improve roadway maintenance sophistication (Kimura 2 August 2013; Daily Engineering and Construction News 9 December 2014).

While 'smart' applies to innovations and initiatives pioneered by market actors, which can be seen to respond to climate change, 'smart' has also been applied in this context to individuals as part of a social engineering campaign to promote road safety. That is, the 'smart driver' (*sumāto doraibā*) campaign pioneered by the Tokyo Smart Driver Project Committee (TSDPC) and Tokyo Smart Rider (TSR), for motorcycle users, both under Metropolitan Expressway Company Limited, among other areas (Aomori Smart Driver 2012; Tochigi Smart Driver 2013; Hyogo Smart Driver 2016; Kumamoto Smart Driver n.d.; Yamanashi Smart Driver 2016), offer guidelines on how to become a 'smart driver' (Tokyo Smart Driver Project Committees 2011a, 2011b; Tokyo Smart Rider 2012). Although it focuses chiefly on avoiding road accidents, like initiatives and innovations, 'smart' individuals can also be seen as agents responding to climate change through market activity and individual behaviour. For example, the private business Recoo, under the Tokyo Smart Driver slogan, aims to allow users to measure how 'smart' one's driving is in terms of fuel consumption, responding to concerns about pollution and climate change, as a component of what it is to be a 'smart driver' (Tokyo Smart Driver Project Committees 2011c, 2011d).

The use of 'smart' within other initiatives under a vision of a 'smart society' similarly reflects this trend, with emphasis on changes to individual behaviours amidst references to initiatives and innovations. Though behavioural change is undoubtedly a necessary component for tackling climate change, it also represents an area for the recalibration of risk between state and individual. For example, recent initiatives for a 'smart platinum society' have centred on responding to Japan's population demographics, where the Ministry of Internal Affairs and Communications (MIC) identifies three main 'visions' (Ministry of Internal Affairs and Communications, Japan 20 June 2014). The first centres on initiatives towards individuals living independently (*jiritsu* – 自立) and self-sufficiently' (*jiritsu* – 自律). The second encourages societal participation and working with a purpose in life, while the third seeks to create new businesses for a hyper-aged society, linking to international markets (Ministry of Internal Affairs and Communications, Japan 20 June 2014).

The emphasis on self-sufficiency relates to notions of 'self-responsibility' and not requiring assistance from the state (Hook and Takeda 2007; see also Pope and Fahey, in progress). To this end, a range of ministries and the Smart Platinum Society Promotion Council (SPSPC) promote 'telework' (*telewāku*) as part of a 'self-supporting model' (*jiei-gata*) of employment (e.g. Ministry of Internal Affairs and Communications, Japan August 2014; Smart Platinum Society Promotion Council July 2014). Telework denotes a form of employment reliant on ICT where workers work from home, a satellite office or 'on the go' (Japan Telework Association n.d.; Nishio 2012: 114). Theoretically, this may facilitate pollution reduction mitigating climate change, a business continuity plan in case of contingencies such as natural disasters or intensified weather events exacerbated by climate change or a greater work–life balance respond-ing to social demands within a hyper-aged society (Nishio 2012). However, there are concerns that the lack of restriction could facilitate a drive for 'higher labour efficiency by allowing labour space to encroach on personal space' on the 'premise that employers can freely confiscate employees' private spaces', while fees for 'at-home work' have been declining drastically in a bid-based system (Sato 2013: 57–59). That is, while telework initiatives may be seen as a contribution to combating climate change, it brings with it a number of other potential problems for the individual.

Despite this, government discourse surrounding telework as a component of smart initiatives typically do not redress these issues. For example, according to the 2014 Smart Platinum Society Promotion Council report (July 2014: 33–34), this new 'workstyle' focuses on centring provision around citizens' needs, similar to smart city designs (Kinjō and Rure 2014: 9). SPSPC identifies the 'needs' of caregivers, women child carers and 'active seniors' (*akutibu shinia*). The needs of the former two are identified as wanting to work whenever they are free amidst caregiving, while the latter as wanting to work utilising their acquired expertise (Smart Platinum Society Promotion Council July 2014: 34). The interests of only these individuals are specifically defined, as opposed to government and business interests of, presumably, increasing productivity and profitability, or worker needs of regular income.[10] Further, in a 'telework guidebook' aimed at helping corporations to implement such initiatives, MLIT, in tandem with a number of other ministries, provide an overview of the efficacies (*kōka*) and utilities (*kōyō*) of telework initiatives to business management, office workers and society, respectively (Ministry of Land, Infrastructure, Transport, and Tourism, Japan n.d.). Telework is seen to benefit businesses in terms of management, productivity, employment flexibility, human resources, image and cost-saving (particularly on work spaces) and the individual in terms of added flexibility to their work–life balance, child safety, choice of residence and cutting down on commuting time. Crucially, for society, it is seen to greatly reduce CO_2 emissions, contributing to 'global warming prevention' (*chikyū ondanka bōshi*). Here based on research conducted by MLIT which suggested telework would reduce the number of those commuting from home to work, it is estimated that it could reduce the nation's annual CO_2 emissions by 3.21 to 4.42 million tonnes (Ministry of Land, Infrastructure, Transport, and Tourism n.d.: 24–25).

In both reports, technological change is seen to enfranchise citizens who require more flexibility and benefit businesses with little to no mention over the extent of encroachment. Further, the SPSPC report advocates new mar-ket conditions based on ICT as a facilitatory mechanism that grants caregiv-ers, child carers and 'active seniors' a means to self-actualise while the MLIT report states that telework will improve 'individuals' self-management abilities' (*kojin-no jiko kanri nōryoku*) (Ministry of Land, Infrastructure, Transport, and Tourism, Japan n.d.: 16). This is mirrored by notions of 'smart work' (*sumāto wāku*) which denote surrogating working long hours for working 'smartly' to increase productivity (Diamond Online 15 January 2015) and dovetails with reintroducing seniors and child carers into the labour force, with an onus on 'independence' (*jiritsu*) and 'self-responsibility' (*jikosekinin*) (Diamond Online 15 January 2015). According to the Japan Association of Corporate Executives, this is carried out by 'smart workers' (*sumāto wākā*) who are 'human resources who independently [*shutai-teki*] choose ways of life and work while possessing a sense of oneself, integrating work into their lives, regardless of position or employment type' (Keizai Doyukai 22 April 2015: 10).

Though not as prominent in government discourse, the state has promoted a number of smart work initiatives. These include the 'Smart Work Challenge 20' in collaboration with SDSK Corporation or 'smart work' awards to major corporations, through a mixture of incentives and disincentives to influence worker practices (Hitachi Solutions Inc 5 November 2014; Cabinet Office Government of Japan March 2015; Development Bank of Japan Inc 21 August 2015: 10; *Nikkei Shimbun* 17 July 2015). Further, it has amended the Labour Standards Law (*rōdō kijun hō*) in 2015, obliging companies to specify the amount of paid leave given, and effectively shifting emphasis onto individual time management (Diamond Online 15 January 2015).

Such discourse is evident in health-based initiatives also. For instance, the smart platinum society vision for healthcare includes a preventive model involving the use of 'network robots',[11] wearable and, potentially, residential sensors for non-contact monitoring, big data analysis through cloud-based electronic health records and the creation of 'life support businesses' (Ministry of Internal Affairs and Communications, Japan 20 June 2014, September 2014). Other projects aim to improve ICT literacy through training sessions, 'self-actualisation' (*jikojitsugen*) through 'learning and teaching together' (Smart Platinum Society Promotion Council July 2014) and the practical use of network human-interface technology (Ministry of Internal Affairs and Communications, Japan 20 June 2014). With this, the conventionally dubbed 'Healthcare Reform Act' (*iryō hōkenseido kaikaku kanren-hō*) enacted in 2015 pushed forward, among other things, 'preventive healthcare' aimed at changing individual behaviour towards healthier lifestyles with a number of incentive measures (Mizuho 2016: 328–330). These include incentives for individuals who carry out healthy practices (Hashimoto 16 October 2014; Asahi Shimbun Company 25 July 2015) while insurers are obligated to support 'self-help efforts' (*jijodoryoku*) among their subscribers, alongside health education, counselling and check-ups, with financial incentives for proactive efforts (Mizuho 2016: 328–329). Moreover, insurers must utilise data on health insurance claims and specific medical examinations to carry out such health projects (Mizuho 2016: 329). Legislative changes have created new duties for healthcare insurers which have manifested into a focus on risk management shifting from 'waiting' to proactively 'catching' and managing illnesses as well as changing subscribers' behaviours, particularly relating to people at risk of lifestyle diseases (Mizuho 2016: 333). The shift in focus effectively engenders behavioural adaptation in the individual to lower overall healthcare costs as tax revenue shrinks due to the effects of a hyper-aged society.

Additionally, it is considered that health promotion can facilitate the emergence of healthcare businesses. For example, the 'Smart Life Project', initiated in 2011

(Ministry of Health, Labour and Welfare, Japan 2015: 436), in the same general model as the 'smart driver' initiatives above, entails a social engineering campaign where private and public actors are incentivised to promote health awareness and hence behavioural change in the individual whether through products, services or other activities. The overall aim is for this increased awareness to percolate throughout society and culminate in a nationwide social movement (Koga 2 September 2014) which also creates a market for new healthcare businesses. Health promotion activities by market actors as well as state actors is recorded on the website of the Smart Life Project Committee (SLPC) (Smart Life Project Committee and Ministry of Health, Labour and Welfare, Japan n.d.) where SLPC and the Ministry of Health, Labour and Welfare, Japan promote four main objectives for citizens[12] regular exercise ('smart walk'), a healthy diet ('smart eat'), to not smoke ('smart breath') and to routinely go to check-ups ('smart check') (Ministry of Health, Labour and Welfare, Japan 2015: 436; Smart Life Project Committee and Ministry of Health, Labour and Welfare, Japan n.d.). The SLPC is chaired by Toshikazu Saitō, chairman of the board of directors of the sports and fitness company Renaissance Inc. (Smart Life Project Committee and Ministry of Health, Labour and Welfare, Japan 11 December 2014), who states in a message posted on the SLPC website:

> Not just 'longevity', but being able to see out every day happily, healthily and with spirit.
> I want to make leading an independent life without being lent a helping hand [lit. borrowing someone's help], in other words, 'healthy longevity', our Japan's goal.
>
> (Saitō 24 February 2014, in Smart Life Project
> Committee and Ministry of Health, Labour and
> Welfare, Japan 11 December 2014)

Here, the notion of collective effort is encapsulated in the conveyance of a national campaign, attributing a role to citizens to live healthily. Also, similar notions of self-sufficiency emerged with 'independence' placed as a central target without 'calling on someone for help', where the agency of those who would help – whether family members, friends, charities/foundations or the state – is left undisclosed. In addition, Saitō adds later on that:

> [. . .] we will develop campaigns in collaboration with companies and organisations. We will support your health so as to be able to revise daily routines [lit. lifestyle cycle] from small day-to-day changes. To aim for Japan to be admired [lit. proud] by the world, as a healthy longevity nation.
>
> (Saitō 24 February 2014 in Smart Life Project
> Committee and Ministry of Health, Labour and
> Welfare, Japan 11 December 2014)

Again, the interests of market actors are not specifically mentioned other than implicitly in making Japan a healthy longevity nation, and they are portrayed as carrying out a facilitating role in the initiative with SLPC. Similarly, it is implied that citizen efforts towards living healthily will see Japan reflected positively in the eyes of the world, while the committee's chief role is supporting the individuals's health, and hence enfranchising the citizen whose active efforts contribute to a new national image.

How preventive healthcare initiatives may affect citizens over time is neither completely clear nor determined. For instance, reports suggest that the government considered the possibility of insurance adjustments for those who do not take certain health check-ups, before amendment (Huffington Post Japan 5 June 2014). Further, amidst a legislative focus on 'incentives', some companies under pressure towards meeting health promotion targets have reportedly formulated, among other initiatives,[13] a 'disincentive system' where employees would have their bonuses cut for not meeting sufficient requirements (Ashitanokenpo Project 2014–15; Collaboherusu kenkyūkai 1 April 2015).

Overall, then, whether for 'smart work', 'smart driving' or 'smart living' in terms of one's health, all initiatives constitute, in part, frameworks for effectuating individual behavioural change. These all denote the improvement of existing frameworks through the use of technology while underscoring the importance of self-reliance and independence in the individual. With this, the use of 'smart' as a buzzword affords a broad scope in terms of its application, allowing external initiatives to thematically link with others and initiatives themselves to transmute in meaning. For example, notions of a 'smart life' have extended beyond responding to a hyper-aged society and relate to individual energy management. The point is clear from the way that Tokyo Electric Power Company (TEPCO) and Chubu Electric Power Company Inc (Chuden) have promoted the 'Smart Life Plan' for different electricity rates (Chubu Electric Power Company Inc n.d.; Tokyo Electric Power Company Inc n.d.). Also, a number of large-scale firms involved in smart initiatives and innovations, such as Daikin Industries, Hitachi Ltd, Fujitsu Ltd, Mitsubishi Electric Corporation, Panasonic Corporation, Sharp Corporation, and Toshiba Corporation (Japan Smart Community Alliance 1 April 2016), are members of the 'Smart Life Forum' (Smart Life Japan 31 March 2016). This is an organisation spearheaded by private actors which targets realising a 'smart life' among Japanese individuals by disseminating the practice of economical consumption of energy. It has also carried out a number of initiatives in collaboration with the Ministry of Economy, Trade and Industry (METI) to promote energy conservation efforts (Kankyō Bijinesu Orain 2 December 2014) as a means to respond to climate change. Similar to the 'smart' campaigns above, it essentially comprises another nationwide social engineering campaign where individual behaviour can drive market creation, which reciprocally reinforces behavioural change. Here market and individual actors are seen to align as implied in their 2015 'Book recommending a smart life':

> The energy conservation and saving campaign is a social movement which tackles energy conservation and saving with families, corporations, the whole of Japan, coming together as a team.
>
> (Smart Life Japan 27 May 2015: 62)

Again, as a national effort, the interests of social and market actors are seen to coalesce where 'families' and 'corporations' work in unison to realise greater energy efficiency. The booklet itself informs consumers on how to save energy using new energy management systems, including smart meters, and various electronic products, and hence live 'smartly'. Here, a smart life is defined as follows:

> A 'smart life' where one manages by 'energy management [lit 'ene-mane']
> (EMS)', combining storage energy devices for things like electric cars, stor-
> age batteries, and energy conversion [lit creation] devices for things like
> fuel cells, solar power, with energy-saving home appliances where power
> consumption has markedly reduced compared to ten years ago.
>
> (Smart Life Japan 27 May 2015: 10)

The aim is to change consumption behaviour in the individual to realise greater energy efficiency through market mechanisms guided and encouraged by the state, responding to climate change. Similar to the roadway and healthcare campaigns above, the role of individual is to proactively utilise market mechanisms as a driving force assisting the national campaign. Here, however, the emphasis is squarely on energy conservation projects and 'smart' consumption of energy by, for example, purchasing energy-efficient products. This increased awareness, again, creates a potential market for new businesses. Further, as with the 'population approach' for preventive healthcare or promoting safe driving, and to a lesser extent, working, the aim essentially is to promote greater awareness to change citizens' living habits and practices towards living 'smartly' – however that may be defined. That is, whether by 'smart time management', taking a 'smart drive' or a 'smart walk' or reading the 'smart meter' – it is an adaptation strategy to minimise risk whether primarily to the individual or the state.

Conclusion

This chapter has provided an exploratory analysis of discourse centred on smart projects looking specifically at the use of 'smart' as a buzzword. While smart initiatives such as those relating to scaling-up the use of renewable energy constitute an exciting means by which to respond to pollution and more general climate change, the wider application of 'smart' technologies raises penetrating questions over the nature of governance in a digital era. Although this section has focused on roles and responsibilities on the societal level, initiatives have centred on changing individual behaviour whether for working, living or commuting, where self-management is key.

The section 'Smart cities and climate change' discussed smart initiatives in Australia, China and Japan, introducing smart initiatives as a global phenomenon and associated with notions of a fourth Industrial Revolution, as ICT is incorporated into infrastructures and public and private services. Section three sought to explore the roles of individuals on the societal level in order to gain more insight into the nature of change brought about by smart initiatives. Further, they were linked to theorisation (Strang and Meyer 1993) insomuch as they relate to innovations affecting new visions of society, life, and, elsewhere, the planet, having potentially huge impacts on social practices. Here, the use of 'smart' as a buzzword obfuscates concrete initiatives as a positively loaded judgement denoting sophistication and style. Further, its usage reflects the 'political qualities' of certain initiatives in encouraging behavioural adaptation, potentially affording greater structural flexibility, but where the agency of those who decide who is 'smart' and who is not remains a point of seminal importance.

In order to explore the potential role of individuals, the qualitative analysis discussed recent initiatives relating to roadways, employment, healthcare and energy use. It was noted that 'smart' as a buzzword is used to describe initiatives, innovations and individuals. Here, discourse was focused on how these new conditions enfranchise citizens, allowing them to 'self-actualise', with similarly themed phraseology observable across a range of initiatives that promote individualism, such as 'self-reliance' and 'self-help' efforts. That is, rather than discussing the potential encroachments into personal space, discourse surrounding initiatives discussed changes in terms of manner, with 'smart work' and, to a larger extent, 'smart life' reflecting new, superior behaviours. The 'smart driver', 'smart rider' and both 'smart life'–derived initiatives denoted efforts towards a nationwide social movement to influence social and individual behaviour and minimise risk, taking place with close collaboration between state and market actors, and in response to different threats to national sustainability. Here, discourse placed 'Japan' or 'society' as the common social category allowing for the dissemination of new practices into the everyday.

As stated from the outset, this chapter does not seek to downplay the necessity of responding to climate change. Rather it aims to explore the 'political qualities' borne out, and not necessarily a prerequisite, of new innovations (Winner 1980). Overall, the use of 'smart' as a buzzword makes it flexible in application and gives it political qualities where 'smart' can apply to the 'three Is' of institutions, initiatives and individuals. For the latter, the use of such ameliorative language serves to encourage citizens to take on additional roles and responsibilities, affording the potential off-loading of state risk onto the market or individual at the societal level. It is within highly abstract visions of a 'smart city', as with other initiatives in the past such as the 'New Life Movement' (Takeda 2005), that individuals are viewed as resources. Here, as states necessarily respond to climate change amidst considerable industrial change it is an important task to assess initiatives and the socialisation process of the practices evoked in terms of how they affect different stakeholders of the future. Although this study was inevitably broad, a point of departure in market-oriented initiatives aimed at responding to, for example, climate change, may be what incentives for those who are 'smart', and what else for those who, fairly or unfairly, do not meet new standards.

Appendix

'Smart' categories

Table 6.1 'Smart' categories compiled from National Diet Proceedings*

Categories	Terms
Agriculture	'agri', agriculture
City	building, city, city planning, community, community city, eco park, factory, home, house, (Okinawa) Island Initiative, platinum, platinum society, residence, region, (compact and [. . .]) region, society, town, town planning, village, wellness residence, wellness residence city, wellness Sanjō City Promotion Plan
Energy facilities	control, demand, demand restraint, energy, energy district, energy management, energy self-sufficiency system, energy supply and demand, energy technology, Energy Week, Energy Week 2014, grid, management, management system, meter, power transmission and distribution network, sensor, system, water
ICT/general	card, economic growth, era, growth, '-ification', 'factorization', 'one stop', policy, technology, ubiquitous net
Individual lifestyle/ consumption behaviour	behaviour, consumption, consumption behaviour, consumption restraint, (Japan [. . .]) Life Institute, Life Project, economical use of energy, energy saving, way of life, wellness
International Relations	acquisition initiative, assistance superpower, diplomacy, diplomatic negotiation, donor, foreign investment, negotiation, ODA, power, regulation, sanctions, state
Products	consumer electronics, phone, TV
Transportation	area, 'change', ic, 'inter', interchange, parking, plate, support station, station, way

(*Continued*)

Table 6.1 (Continued)

Categories	Terms
Other	administrative organization, advert, answer, appearance, Basic Law on Fisheries, 'biz', body, ceremony, chip, cruise, economic and fiscal policy, expression, factory, fashion, figure/form/shape, image, island, law system, material, method, minister, name, naming, open service, person, power plant, policy, public office, pose, provision (act, treaty), question, shelf, tournament, thing, thought process, weapon

Source: Author[14]

* There are number of issues concerning how these concepts were categorised. For example, many of the terms within the City category such as 'home', 'house' and 'residence' might only be considered 'smart' due to the upgraded energy facilities installed and so should logically be placed in the Energy category. However, though the 'smart city' concept is closely related to the ways in which the energy infrastructure that undergirds it is being redesigned, it also potentially constitutes an integration of ICT to other urban infrastructures to develop a new business model for economic growth (Ministry of Economy, Travel and Industry, Japan 24 April 2014). As a result, the 'Smart City' category draws a line between 'energy facilities' as well as other smart projects such as the 'smart interchange', and general references regarding 'smart ICT' and 'smart growth', to include terms that denote a 'living space', from the 'smart home' to 'community', 'city' and 'society'. Similarly, there is a strong case to be made that a 'smart person' might be considered as belonging to the category 'Individual Lifestyle/Consumption Behaviour' (see Vanolo 2014), but references to people were categorised as 'Other' as it is possible that they are not related to 'smart' in this context.

Notes

1 See also Goatly (2007) for a related but more general discussion on ideological naturalisation based on the use of figurative and conventionalised language.
2 Unless otherwise stated the translations in this chapter are the author's.
3 Denoting 'Internet of Everything'.
4 Which is also involved in the Asian Super Grid (Mathews 2012).
5 This is also considered to extend to receptivity of measures relating to smart grids (see DeWit 2014a).
6 This is following the research by Wakuta (2015) which demonstrated that the Japanese legislature is a useful point of departure for assessing processes of dissemination and institutionalisation of new practices in the country in line with theorisation.
7 The 'International Relations' category applies to the efficacious use of diplomacy and 'smart power' (Armitage and Nye 2007), denoting, very loosely, the projection of (hard and soft) power 'in ways that are cost-effective and have political and social legitimacy' (Crocker *et al.* 2007: 13).
8 Denoted as a 'car that does not crash' (*butsukaranai kuruma*) (Highway Industry Development Organization 2011b: 3).
9 The 'Gas Business Act' was also amended to liberalise the gas retail market in 2017 allowing new companies to compete with major city gas firms in large cities (Nikkei Shimbun 17 June 2015; *The Japan Times* 5 July 2015).
10 Though increased labour flexibility theoretically need not preclude regular income.
11 A network robot refers to a robot linked to a ubiquitous network allowing users to download new applications/functions into a purchased robot to meet a given

demand such as education, employment, environment, healthcare, industry and transportation, among others (see Nippon Denshin Denwa 2005).

12 Many other localised initiatives are orchestrated round these objectives, but, in the case of the Kumamoto Prefectural Government, for example, may add more such as 'smart oral care' (promoting oral hygiene) and 'smart refresh' (promoting sufficient rest) (Kumamoto Prefecture Government 22 March 2016).

13 For example, 'smart life stay' for high-risk employees. 'Smart life stay' comprises a mixture of public and private initiatives to provide visits to heath education resorts involving specialist medical advice and education over diet and exercise alongside activities (Nihon Iryō Kenkyū Kaihatsu Kikō 15 September 2015).

14 The term 'SMArt 155' ($n = 1$), denoting a German artillery round, was excised from the list and the phrase 'smart 2.5billion yen' ($n = 1$) was omitted due to potential ambiguity.

References

Akiyama, Takashi (15 July 2006) 'Kokukōshō, kōzokudōro no 'sumāto IC', dōnyū o honkaku-ka suru to happyō' [MLIT announce the introduction of 'smart IC' expressways]. *SC-Abeam Automotive Consulting*. Accessed on 4 March 2016 from www.sc-abeam.com/sc/?p=1592

Andrews, Rhys, and Steven van de Walle (2012) 'New public management and citizens: Perceptions of local service efficiency, responsiveness, equity and effectiveness'. *Coordinating for Cohesion in the Public Sector of the Future*, COCOPS Working Paper No. 7. Accessed on 9 February 2016 from www.cocops.eu/wp-content/uploads/2012/08/COCOPS_workingpaper_No7-.pdf

Aomori Smart Driver (2012) 'aomori sumāto doraibā-ni tsuite' [About Aomori Smart Driver]. *Aomori Smart Driver Project Committees*. Accessed on 27 April 2016 from www.aomori-smart-driver.com/about

Armitage, Richard Lee, and Joseph S. Nye, Jr. (2007) *CSIS Commission on Smart Power: A Smarter, More Secure America*. Washington, DC: Center for Strategic and International Studies.

Arup (2014) 'Smart grid, smart city: Shaping Australia's energy future–Executive report'. *Ausgrid*. July 2014. Accessed on 9 February 2016 from www.industry.gov.au/Energy/Programmes/SmartGridSmartCity/Documents/SGSC-Executive-Report.pdf

Asahi Shimbun Company (25 February 1998) *Kaden ya jidōsha no kōritsu kijun o kyōka shōene-hō kaisei-an* [Strengthening of home appliances and automobile energy efficiency standards–Amendment to the Energy Conservation Act].

Asahi Shimbun Company (22 September 2006) *Kōsoku-dō sumāto IC, 18-kasho-o jōsetsu-ni* [Smart IC expressways, 18 locations to become permanent].

Asahi Shimbun Company (4 March 2009) *(Bijinesu yōgo jiten enerugī-hen) shōene-hō'* [(Business glossary: Energy edition) The save-energy law].

Asahi Shimbun Company (18 June 2015) *Denki gasu, katei ga erabu kaisei-den jihō nado seiritsu, denryoku jiyū-ka* [Households choose electricity and gas–Establishment of amended Electricity Business Act, liberalization of electric power]. Accessed on 10 March 2016 from www.asahi.com/articles/DA3S11812791.html

Asahi Shimbun Company (25 July 2015) *Yobō de iryō-hi yokusei/ haikō o kashidashi seiken, aratana saishutsu kaikaku* [Medical cost control through prevention/lending out closed school–the Administration's new expenditure reform].

Ashitanokenpo Project (2014–15) 'Kigyō kenpo hōmon shirīzu kabushikigaisha rōson' [Company/Health Insurance vist series: Lawson, Inc.]. *Fujio Akatsuka*. Accessed on 17 March 2016 from www.ashiken-p.jp/column/08.html

Australian Communications and Media Authority (2015) 'The Internet of Things and the ACMA's areas of focus: Emerging issues in media and communications'. *November 2015 Occasional Paper.* Canberra: Commonwealth of Australia. Accessed on 9 February 2016 from www.acma.gov.au/theacma/internet-of-things-and-the-acmas-areas-of-focus-occasional-paper

Bajracharya, Bhishna, David Cattell, and Isara Khanjansthiti (2014) 'Challenges and Opportunities to Develop a Smart City: A Case Study of the Gold Coast, Australia'. Paper presented at *Real CORP 2014: Plan it Smart, Vienna, Austria.*

Cabinet Office Government of Japan (March 2015) *shanai-ni okeru wāku raifu baransu shintō/teichaku-ni muketa pointo/kōjireishū* [Instances and good case studies towards the permeation and establishment of a work-life balance within company]. Accessed on 30 April 2016 from hwwwa.cao.go.jp/wlb/research/wlb_h2703/chapter4.pdf

Cabinet Office Government of Japan (19 March 2015) 'mirai-no sangyō sōzō/shakai henkaku-ni muketa torikumi (soan) (dai 4-ji sangyō kakumei-ni yoru chōsumātoshakai (kashō)-no kōchiku)' [Initiative (draft) towards the creation of future industries and social change (The construction of the Super Smart Society (provisional name) due to 4th Industrial Revolution]. *Material for the 4th Basic Plan Executive Committee from Scientific Technology and Innovation Conference.* Accessed on 30 March 2016 from www8.cao.go.jp/cstp/tyousakai/kihon5/4kai/siryo4-1.pdf

Cabinet Office Government of Japan (22 January 2016) *kagaku gijutsu kihon keikaku* [Science and Technology Basic Plan]. Accessed on 12 May 2016 from www8.cao.go.jp/cstp/kihonkeikaku/5honbun.pdf

Chen Weiwei (7 September 2015) 'Guojia nengyuan ju juzhang Nu Er·Baikeli jiedu "pei dian wang jianshe gaizao xingdong jihua (2015–2020 nian)"' [An interpretation of National Energy Administration secretary, Nur Bekri: 'The Distributed Network Construction and Reform Action Plan (2015–2020)]. *Zhongguo zhengfu wang.* Accessed on 13 May 2016 from www.gov.cn/zhengce/2015–09/07/content_2926444.htm

China Market Research (5 May 2016) 'Smart electric meter industry analysis 2016 and market outlook 2020 for China'. *Newsgallery.* Accessed on 13 May 2016 from www.newsmaker.com.au/news/53584/smart-electric-meter-industry-analysis-2016-and-market-outlook-2020-for-china#.VzR8QrzIbR4

Chu, Cheng-Kang, Joseph K. Liu, Jun Wen Wong, Yunlei Zhao, and Jianying Zhou (2013) 'Privacy-preserving smart metering with regional statistics and personal enquiry services'. *ASIA CCS 2013: Proceedings of the 8th ACM SIGSAC Symposium on Information, Computer and Communications Security,* 369–379. doi:10.1145/2484313.2484362.

Chubu Electric Power Company Inc (26 November 2013) *sumātomētā-no dōnyū keikaku* [Smart meter introduction plan]. Accessed on 10 March 2016 from www.chuden.co.jp/corporate/publicity/pub_release/press/3234543_6926.html

Chubu Electric Power Company Inc (15 April 2014) *Shōenehō kaisei-ni moto-dzuku sumātomētā dōnyūnado-ni kansuru keikaku-no kōhyō-ni tsuite* [On the public announcement of plan relating to the introduction of smart meters based on the amendement to the Energy Conservation Act, etc.]. Accessed on 10 March 2016 from www.chuden.co.jp/corporate/publicity/pub_oshirase/topics/3239636_19351.html.

Chubu Electric Power Company Inc (n.d.) *sumāto raifu puran* [Smart life plan]. Accessed on 4 April 2016 from www.chuden.co.jp/business/bshikumi/business_menu/bus_shop/dento/smart/index.html

Chung, Eui-Seok, Brian Soden, B. J. Sohn and Lei Shi (2014) 'Upper-tropospheric moistening in response to anthropogenic warming,' *Proceedings of the National Academy of Sciences of the United States of America* 111(32): 11636–11641.

Collaboherusu kenkyūkai (1 April 2015) 'goaisatsu' [Greetings]. *EWEL, Inc.* Accessed on 1 April 2016 from www.collabo-health.jp/index.html

Cormack, Lucy (24 February 2016) 'Smart meters rolling out in NSW to "change the way we think about electricity"'. *The Sydney Morning Herald.* Accessed on 8 March 2016 from www.smh.com.au/business/consumer-affairs/smart-meters-rolling-out-in-nsw-to-change-the-way-we-think-about-electricity-20160224-gn2jb3.html

Cornwall, Andrea (2007) 'Buzzwords and fuzzwords: Deconstructing development discourse'. *Development in Practice* 17(4/5): 471–484.

Cornwall, Andrea (2010) 'Introductory overview – buzzwords and fuzzwords: Deconstructing development discourse', in Andrea Cornwall and Deborah Eade (eds) *Deconstructing Development Discourse: Buzzwords and Fuzzwords.* Warwickshire, UK: Oxfam and Practical Action Publishing, pp. 1–18. Accessed on 25 March 2016 from www.guystanding.com/files/documents/Deconstructing-development-buzzwords.pdf

Cornwall, Andrea and Karen Brock (2005) 'What do buzzwords do for development policy? A critical look at "Participation," "Empowerment," and "Poverty Reduction"'. *Third World Quarterly* 26(7): 1043–1060.

Cornwall, Andrea and Deborah Eade (eds) (2010) *Deconstructing Development Discourse: Buzzwords and Fuzzwords.* Warwickshire, UK: Oxfam and Practical Action Publishing.

Crocker, Chester A., Fen Osler Hampson, and Pamela Aall (2007) 'Introduction: Leashing the dogs of war', in Chester A. Crocker, Fen Osler Hampson, and Pamela Aall (eds) *Leashing the Dogs of War: Conflict Management in a Divided World.* Washington, DC: United States Institute of Peace Press, pp. 3–16.

Daily Engineering & Construction News (9 December 2014) 'higashi nihon kōsoku geisha/saitama-shini shinkansei sentā, 16-nen haru kaisetsu/bōsai kinō kyōka' [NEXCO East to set up new control centre in Saitama City, to be opened in Spring 2016 – strengthening disaster prevention functions]. *The Nikkan Kensetsu Kogyo Shinbun.* Accessed on 27 April 2016 from www.decn.co.jp/?p=20273

Darby, Sarah (2006) *The Effectiveness of Feedback on Energy Consumption: A Review for DEFRA of the Literature on Metering, Billing and Direct Displays.* Environmental Change Institute, University of Oxford. Accessed on 9 March 2016 from www.usclcorp.com/news/DEFRA-report-with-appendix.pdf

Darby, Sarah (2013) 'The role of smart meters in carbon management'. *Carbon Management* 4(2): 111–113.

Department for Energy and Climate Change (2010) '2050 Pathways analysis'. *Crown copyright.* Accessed on 9 March 2016 from www.gov.uk/government/uploads/system/uploads/attachment_data/file/68816/216-2050-pathways-analysis-report.pdf

Development Bank of Japan Inc (21 August 2015) 'wāku raifu baransu-no jitsugen-ni yoru rōdōryoku kakuho/seisansei-no kojō-ni mukete' [Towards improvement in productivity and ensuring the labour force through the realization of work-life balance]. *kongetsu-no topikkusu* 239(1). Accessed on 30 April 2016 from www. dbj.jp/pdf/investigate/mo_report/0000020311_file4.pdf

DeWit, Andrew (2011) 'Fallout from the Fukushima shock: Japan's emerging energy policy'. *The Asia-Pacific Journal* 9(45): Number 5. Accessed on 7 February 2016 from http://apjjf.org/2011/9/45/Andrew-DeWit/3645/article.html

DeWit, Andrew (2012) 'Distributed power and incentives in Post-Fukushima Japan'. *The Asia-Pacific Journal* 10(49): Number 2. Accessed on 3 March 2016 from http://apjjf.org/2012/10/49/Andrew-DeWit/3861/article.html

DeWit, Andrew (2014a) 'Japan's rollout of smart cities: What role for the citizens?'. *The Asia-Pacific Journal* 11(24): Number 2. Accessed on 29 February 2016 from http://apjjf.org/2014/11/24/Andrew-DeWit/4131/article.html

DeWit, Andrew (2014b) 'A new Japanese miracle? Its hamstrung feed-in tariff actually works'. *The Asia-Pacific Journal* 21(38): Number 2. Accessed on 11 February 2016 from http://apjjf.org/2014/12/38/Andrew-DeWit/4185/article.html

DeWit, Andrew (2015) 'Japan's dangerous nuclear waste on the cutting board? Towards a renewables future'. *The Asia-Pacific Journal* 13(44): Number 2. Accessed on 15 January 2016 from http://apjjf.org/-Andrew-DeWit/4396

Diamond Online (15 January 2015) '"2015-nen no hatarakikata" "Muda-na chōjikan kinmu"-kara "sumātowāku"-he henka-no kagi-to naru-nowa kojin-no ishiki kaikaku — ōkubo yukio rikurūtowākusu kenkyūjochō-ni kiku' [Way to work in 2015: The key to changing from 'wasteful long working hours' to 'smart work' is a change in individuals' mentality – An interview of Yukio Okubo, head of the Recruit Works Research Institute]. *Diamond Inc.* Accessed on 10 August 2016 from http://diamond.jp/articles/-/64970

Dirks, Susanne, Constantin Gurdgiev, and Mary Keeling (2010) 'Smarter cities for smarter growth: How cities can optimize their systems for the talent-based economy'. *IBM Institute for Business Value*, IBM Global Business Services Executive Report. Accessed on 4 February 2016 from http://papers.ssrn.com/sol3/papers.cfm?abstract_id=2001907

Doyle, John, Andrew Evans, Verena Juebner, and Jennifer Chan (2015) 'Realising the benefits of smart meters'. *Victorian Auditor-General's Report*, September, Victorian Auditor-General's Office. Accessed on 9 March 2016 from www.audit.vic.gov.au/publications/20150916-Smart-Meters/20150916-Smart-Meters.html

EandE Solutions Inc (1 August 2012) 'Saisei kanō enerugī tokubetsu sochi-hō'-ga shikō saremashita sono 1' [The "Renewable Energy Special Measures Act" went into effect. Number 1]. DOWA ekojānaru, *DOWA Eco-System Co., Ltd.* Accessed on 11 March 2016 from www.dowa-ecoj.jp/houki/2012/20120801.html

e-Gov (n.d.) 'Denki jigyō-sha ni yoru saisei kanō enerugī denki no chōtatsu ni kansuru tokubetsu sochi-hō' [Act on special measures concerning renewable energy electric procurement by operator of electric utilities]. *Ministry of Internal Affairs and Communications.* Accessed on 11 March 2016 on http://law.e-gov.go.jp/htmldata/H23/H23HO108.html

Energia (April 2014) 'sumātomētānado-no seibi-ni kan suru keikaku' [Plan regarding the development of smart meters etc.]. *The Chugoku Electric Power Co., Inc.* Accessed on 10 March 2016 from www.energia.co.jp/energy/energia/energy.pdf

Giffinger, Rudolf, Christian Fertner, Hans Kramar, Robert Kalasek, Nataša Pichler-Milanović, and Evert Meijers (2007) *Smart Cities – Ranking of European Medium-Sized Cities.* Vienna, Austria: Centre of Regional Science, Vienna

University of Technology. Accessed on 18 January 2016 from www.smart-cities. eu/download/smart_cities_final_report.pdf

Go, Woong, SeulKi Choi, and Jin Kwak (2013) 'Privacy protection based secure data transaction protocol for smart sensor meter in smart grid'. *International Journal of Distributed Sensor Networks*: 1–9. Accessed on 9 March 2016 from www.hindawi. com/journals/ijdsn/2013/829721

Goatly, Andrew (2007) *Washing the Brain–Metaphor and Hidden Ideology*. Amsterdam: John Benjamins Publishing Company.

Goo Jisho (n.d.) 'Sumāto [smart]' [Smart [Smart]]. *NTT Resonant Inc.* Accessed on 11 March 2016 from http://dictionary.goo.ne.jp/jn/119920/meaning/m0u/%E3%82%B9%E3%83%9E%E3%83%BC%E3%83%88

Gopakumar, Pathirikkat, M. Jaya bharata Reddy, and Dusmanta Kumar Mohanta (2014) 'Letter to the editor: Stability concerns in smart grid with emerging renewable energy technologies'. *Electric Power Components and Systems* 42(3–4): 418–425. doi:10.1080/15325008.2013.866182

Haas, Peter M. (1992) 'Introduction: Epistemic communities and international policy coordination'. *International Organization* 46(1): 1–35.

Hanley, Steve (18 September 2015) 'Japan pushes forward with hydrogen society ahead of Olympics'. Gas2, *Sustainable Enterprises Media, Inc.* Accessed on 23 January 2016 from http://gas2.org/2015/09/18/japan-pushes-forward-hydrogen-society-ahead-olympics/

Hanshin Expressway Company Limited (n.d.) *ETC Electronic Toll Collection System.* Accessed on 8 March 2016 from www.hanshin-exp.co.jp/drivers/etc/

Harlan, Chico (4 June 2013) 'In Japan, new policy spurs solar panel boom'. *The Washington Post*. Accessed on 11 March 2016 from www.washingtonpost.com/world/asia_pacific/in-japan-new-policy-spurs-solar-power-boom/2013/06/04/63ce9556-c9cf-11e2-9245-773c0123c027_story.html

Hashimoto, Yoshiko (16 October 2014) '"Seikatsu kaizen de genkin kyūfu". hokensha mo shiji sezu iryō-hi tekisei-ka, hokensha ni wa insentibu' ['Cash benefits for lifestyle improvement' without even insurer's support: Optimization of health care expenses – incentives to insurers]. *M3 Inc.* Accessed on 21 March 2016 from www. m3.com/open/iryoIshin/article/260860

Hayashi, Yoshitsugu (2011) 'Rejiliento-na kokudo-to shaki-ni muketa sumāto shurinku-no susume' [Recommendation for smart shrink towards a resilient national infrastructure and society]. *IATSS Review* 36(2): 139–143.

Heather, Ben (12 February 2015) 'Smart meters: Power companies know when you're home'. *Stuff*. Accessed on 1 March 2016 from www.stuff.co.nz/technology/gadgets/66079055/smart-meters-power-companies-know-when-youre-home

Highway Industry Development Organization (2011a) 'Kōzokudōro-o katsuyō shita chiiki-no kassei-ka bukai-no hōkoku gaiyō: shin-dōro rikatsuyō kenkyūkai' [Overview of report on the Local Revitalization Committee that utilized high-speed roads: Practical use and application of new roads Research Society]. *Dōro gyōsei seminā*, 2011.10. Accessed on 7 March 2016 from www.hido.or.jp/14gyousei_bac knumber/2011data/1110/1110kenkyukai_chiiki_kasseika_HIDO.pdf

Highway Industry Development Organization (2011b) *ITS Handbook*. Accessed on 27 April 2016 from www.hido.or.jp/itsapq/jsp/auth/01_ITShandbook2011_R.pdf

Highway Industry Development Organization (August 2014) *7. kankyō fuka teigen* [7. Reducing environmental impact]. Accessed on 11 August 2016 from www. hido.or.jp/study/files/pdf/application_06_7.pdf

Hirai, Setsuo, Masuo Kawana, Hiroyuki Oouchi, and Yasuyuki Manabe (2005) '3138 The function of ETC roadside system for smart interchange/parking area connection type as a social experiment'. *National Institute for Land and Infrastructure Management*. Accessed on 7 March 2016 from www.nilim.go.jp/lab/qcg/english/3paper/pdf/2005_11_itswc_5.pdf

Hitachi Solutions Inc (5 November 2014) 'Hitachi soryūshonzu-ga 'ikumen kigyō awādo 2014' tokubetsu shōreishō-wo jushō' [Hitachi Solutions received the special award 'Ikumen Companies Award 2014']. *News Release*. Accessed on 30 April 2016 from www.hitachi-solutions.co.jp/company/press/news/2014/1105.html

Hitt, Jack (20 April 2016) 'The future of cars is already here'. *The Smithsonian Institution*. Accessed on 27 April 2016 from www.smithsonianmag.com/innovation/future-cars-already-here-180958777/?no-ist

Hokkaido Electric Power Company Inc (10 April 2014) 'sumātomētānado-no dōnyūkeikaku-ni tsuite' [On the smart meter introduction plans]. *HEPCO Hokkaido Electric Power Co., Inc.* Accessed on 10 March 2016 from www.hepco.co.jp/info/info2014/__icsFiles/afieldfile/2014/04/10/140410.pdf

Honshu-Shikoku Bridge Expressway Company Limited (2005) *ETC Jōhō* [ETC information]. Accessed on 8 March 2016 from www.jb-honshi.co.jp/customer_index/etc/

Hook, Glenn D. (2012) 'Recalibrating risk and governing the Japanese population'. *Critical Asian Studies* 44(2): 309–327.

Hook, Glenn D., Ra Mason, and Paul O'Shea (2015) *Regional Risk and Security in Japan: Whither the Everyday*. Sheffield Center for Japanese Studies/Routledge Series. London: Routledge Taylor and Francis Group.

Hook, Glenn D. and Hiroko Takeda (2007) '"Self-responsibility" and the nature of the postwar Japanese state: Risk through the looking glass'. *Journal of Japanese Studies* 33(1): 93–123.

Huffington Post Japan (5 June 2014) *kenkōshindan-wo jushin shinaito hokenryō hikiage-mo seifu-ga kentō* [Government consdering even increasing healthcare premiums when you don't go for medical checkups]. Accessed on 11 March 2016 from www.huffingtonpost.jp/2014/06/05/health-checkup_n_5450844.html

Hughes, Nick, Neil Strachan, and Robert Gross (2013) 'The structure of uncertainty in future low carbon pathways'. *Energy Policy* 52: 45–54.

Hyogo Smart Driver (2016) *hyōgo sumāto doraibā-ni tsuite* [About Hyogo Smart Driver]. Accessed on 27 April 2016 from http://hyogo-smartdriver.net/about.html

Ikeda, Saburo (2013) 'Beyond conventional scope of risk analysis: Lessons from the 3.11 earthquake, tsunami, and Fukushima nuclear disaster', in Saburo Ikeda and Yasunobu Maeda (eds) *Emerging Issues Learned from the 3.11 Disaster as Multiple Events of Earthquake, Tsunami and Fukushima Nuclear Accident*. The Committee of the Great East Japan Disaster, Society for Risk Analysis, SRA Japan, pp. 15–20. Accessed on 10 January 2016 from www.sra-japan.jp/cms/uploads/311Booklet.pdf

Ishida, Masaya (1 September 2015) 'enerugībunya-de 9757okuen, 2016nendo-no gaisanyōkyū-ha shōene-ni jūten' [Field of energy at 975.7billion yen; budgetary request for 2016 stress energy conservation]. Smart Japan, *ITmedia Inc.* Accessed on 20 January 2016 from www.itmedia.co.jp/smartjapan/articles/1509/01/news049.html

Ishida, Masaya (27 January 2016) '"Chō sumāto shakai"-wo kokka senryaku-de jitsugen-he, enerugībaryūchēn-wo saiteki-ka' [Towards the realization of 'Super

Smart Society' as a national strategy–Optimization of the energy value chain].
Smart Japan, *Itmedia Inc.* Accessed on 23 March 2016 from www.itmedia.co.jp/
smartjapan/articles/1601/27/news036.html

Jagers, Sverker C. and Johannes Stripple (2003) 'Climate governance beyond the
state'. *Global Governance* 9(3): 385–399.

Japan Expressway Holding and Debt Repayment Agency (2014) *Japan Expressway
Holding and Debt Repayment Agency.* Accessed on 7 March 2016 from www.
jehdra.go.jp/english/pdf/others/117.pdf

Japan Smart Community Alliance (1 April 2016) *Japan Smart Community Alliance.*
Accessed on 4 April 2016 from www.smart-japan.org/english/memberslist/index.
html

Japan Telework Association (n.d.) *Terewāku-no dōnyū katsuyō-ni mukete* [Towards
the introduction and application of telework]. Accessed on 2 April 2016 from
www.japan-telework.or.jp/intro/tw_about.html

Japan Times (5 July 2015) 'Electricity and gas liberalization'. Accessed on 5 February
2016 from www.japantimes.co.jp/opinion/2015/07/05/editorials/electricity-
and-gas-liberalization/#.VrSz8vHDNFU

Japan Times (2 October 2015) 'Expressway tolls amid privatization'. Accessed
on 4 March 2016 from www.japantimes.co.jp/opinion/2015/10/02/editorials/
expressway-tolls-amid-privatization/#.VuqZQRjDNFV

Kankyō Bijinesu Onrain (2012) 'kankyō yōgo-shū: Shōene-hō (kaisei shōene-hō)'
[Environment glossary: Energy Conservation Act (Amended Energy Conservation
Act)]. *Nihon bijinesu shuppan.* Accessed on 10 March 2016 from www.kankyo-
business.jp/dictionary/000183.php

Kankyō Bijinesu Onrain (2 December 2014) 'Keisanshō 'fuyu-no setsuden/
shōene kyanpēn' chūsen-de danbōkigu-ya shōhinken purezento' [METI's 'winter
energy-saving/conservation campaign–Heating equipment and gift certificate raffle
prizes]. *Nihon bijinesu shuppan.* Accessed on 5 April 2016 from www.kankyo-
business.jp/news/009291.php

Kankyō Bijinesu Onrain (2 September 2015) 'kankyō-no 2016nendo yosan,
enerugīkanren-ha 60%zō chiiki-no shōene/saiene oshi' [Ministry of Environment
2016 budget, 62per cent in energy, promotion of regional energy conservation
and renewable energy]. *Nihon bijinesu shuppan.* Accessed on 23 January 2016
from www.kankyo-business.jp/news/011232.php

Kansai Electric Power Company Inc (KEPCO) (17 March 2014) 'teiatsujuden-no
okyakusama-he no sumātomētā dōnyū keikaku-no minaoshi oyobi kōatsujuden (500kW
miman)-no okyakusama-muke sumātomētāhe no tsūshinkiki-no dōnyū keikaku-ni
tsuite – zenkoku-ni sakigake, subete-no okyakusama-ni sumātomētā-o katsuyō shita
sābisu-o teikyō' [Plan to introduce communication equipment for smart meters for
customers with high-voltage power (less than 500kW) as well as a review of the plan to
introduce smart meters to customers with low-voltage power]. *Press Release.* Accessed
on 10 March 2016 from www.kepco.co.jp/corporate/pr/2014/0317_1j.html

Kantei (14 June 2013) 'Japan Revitalization Strategy: Japan is back'. *Provisional,
Cabinet Public Relations Office Cabinet Secretariat.* Accessed on 2 March 2016
from www.kantei.go.jp/jp/singi/keizaisaisei/pdf/en_saikou_jpn_hon.pdf

Kasahara, Shouhei and Toshihiko Iwamoto (2015) 'Toshi kōzō-no kaikaku-to
shūyaku-gata toshi-no jitsugen' [Reform of urban structure and implementation
of intensive city]. *Tōkyōjōhōdaigaku kenkyū ronshū* 19(1): 49–58.

Keizai Doyukai (22 April 2015) 'Sekai-ni tsūzuru hataraki-kata ni kan-suru kigyō keieisha-no kōdō sengen – shutai-tekina kojin ni yoru sumāto wāku no jitsugen o mezashite' [Business managers' action declaration related to work style open to the world: Towards the realization of smart work through independent individuals]. *Keizai Doyukai.* Accessed on 28 April 2016 from www.doyukai.or.jp/policyproposals/articles/2015/pdf/150422a.pdf

Kimura, Hayao (2 August 2013) 'higashi nihon kōsoku-ga ICT katsuyō, iji kanri-wo kōritsu/ kōdoka' [East Japan Expressways increasing sophistication and efficiency of maintenance management, using ICT]. Nihon Keizai Shimbun. *Nikkei Inc.* Accessed on 27 April 2016 from www.nikkei.com/article/DGXNASFK0201H_S3A800C1000000/

Kingston, Jeff (2012) 'Mismanaging risk and the Fukushima nuclear crisis'. *The Asia-Pacific Journal* 10(12): Number 4. Accessed on 23 January 2016 from http://apjjf.org/2012/10/12/Jeff-Kingston/3724/article.html

Kinjō, Nanae and Mikako Rure (May 2014) 'The Smart City: An inquiry on the issue involved in going from trial to deployment – Towards the formation of the smart city which produces a citizen-centered "precious circle"' [sumātoshiti: jisshō-kara jigyōka-ni muketa kadai-no ichikōsatsu – shiminchūshin-de 'pureshasu sākuru'-wo umu sumātoshiti-no kēsē-ni mukete. *Ernst & Young Institute Japan.* Accessed on 5 February 2016 from http://eyi.eyjapan.jp/knowledge/future-society-and-industry/2014-05-23.html

Kojima, Megumi (24 March 2012) 'Act on special measures concerning renewable energy electric procurement by operators of electric utilities'. *Waseda University Institute of Comparative Law.* Accessed on 11 March 2016 from www.waseda.jp/hiken/en/jalaw_inf/topics2011/013kozima.html

Kudo, Hiroko (2004) '7. Reform of public management through ICT: Interface, accountability and transparency', in Lawrence Jones, Kuno Schedler, and Riccardo Mussari (eds) *Research in Public Policy Analysis and Management: Strategies for Public Management Reform* (Book Series, Vol. 13). Bingley, UK: Emerald Group Publishing Limited, pp. 153–174.

Kumamoto Prefecture Government (22 March 2016) *Kumamoto sumāto raifu purojekuto ōen dan* [Kumamoto Smart Life Project support groups]. Accessed on 4 April 2016 from www.pref.kumamoto.jp/kiji_5970.html

Kumamoto Smart Driver (n.d.) *kumamoto sumāto doraibā* [Kumamoto Smart Driver]. Accessed on 27 April 2016 from http://kuma-smartdriver.sakura.ne.jp/?page_id=40

Kushida, Kenji E. (2011) 'Leading without followers: How politics and market dynamics trapped innovations in Japan's domestic "Galapagos" telecommunications sector'. *Journal of Industry, Competition and Trade* 11(3): 279–307. doi:10.1007/s10842-011-0104-7

Kushida, Kenji E. (2015) 'The politics of commoditization in global ICT industries: A political economy explanation of the rise of Apple, Google, and industry disruptors'. *Journal of Industry, Competition and Trade* 15: 49–67.

Kyuden (30 April, 2014) 'shōenehō kaisei-ni motodzuku sumātomētā dōnyūnado-ni kan suru keikaku-ni tsuite' [On the plan regarding the introduction of smart meters based on the amendment to the Energy Conservation Act, etc.]. Press Release, *Kyshu Electric Power Co., Inc.* Accessed on 10 March 2016 from www.kyuden.co.jp/notice_140430.html

Lovins, Amory B. (8 July 2014) 'How opposite energy policies turned the Fukushima disaster into a loss for Japan and a win for Germany'. *Rocky Mountain Institute.* Accessed on 4 February 2016 from http://blog.rmi.org/blog_2014_07_08_opposite_energy_policies_turned_fukushima_disaster_into_a_loss_for_japan_and_a_win_for_germany

MacDonald, Mott (2007) *Appraisal of Costs & Benefits of Smart Meter Roll Out Options.* Accessed on 1 May 2016 from http://webarchive.nationalarchives.gov.uk/+/www.berr.gov.uk/files/file45997.pdf

Mason, Ra (2012) *Japan's Recalibration of Risk: The Framing of North Korea.* PhD Dissertation, University of Sheffield.

Mason, Ra (2014) *Japan's Relations with North Korea and the Recalibration of Risk.* Oxon, UK: Routledge.

Mathews, John A. (2012) 'The Asian super grid'. *The Asia-Pacific Journal* 10(48): Number 1. Accessed on 8 February 2016 from http://apjjf.org/-John_A_-Mathews/3858

Mathews, John A. and Hao Tan (2014) 'China's continuing renewable energy revolution: Global implications'. *The Asia-Pacific Journal* 12(12): Number 3. Accessed on 8 February 2016 from http://apjjf.org/2014/12/44/John-A.-Mathews/4209.html

McDaniel, Patrick and Stephen McLaughlin (May-June 2009) 'Security and privacy challenges in the smart grid'. *IEEE Security and Privacy* 7(3): 75–77.

McKenna, Eoghan, Ian Richardson, and Murray Thomson (2012) 'Smart meter data: Balancing consumer privacy concerns with legitimate application'. *Energy Policy* 41: 807–814.

Meinshausen, Malte, Nicolai Meinshausen, William Hare, Sarah C. B. Raper, Katja Frieler, Reto Knutti, David J. Frame, and Myles R. Allen (2009) 'Greenhouse-gas emission targets for limiting global warming to 2°C'. *Nature* 458: 1158–1162.

Metropolitan Expressway Company Limited (n.d.) *ETC-ni tsuite* [About ETC]. Accessed on 8 March 2016 from www.shutoko.jp/fee/etc/

Ministry of Economy, Trade and Industry, Japan (24 April 2014) 'Wagakuni-no sumātokomyuniti jigyō-no genjō: gaiyō' [The current state of our countries' smart community businesses: An overview]. *Energy Conservation and New Energy Department, Agency for Natural Resources and Energy.* Accessed on 2 March 2016 from www.meti.go.jp/committee/summary/0004633/pdf/016_02_00.pdf

Ministry of Economy, Trade and Industry, Japan (August 2015) *Heisei 28-nendo shigen enerugī kankei gaisan yōkyū no gaiyō* [Fiscal year 2016 – Budgetary appropriations related to resources and energy: Overview of request]. Accessed on 3 February 2016 from www.meti.go.jp/main/yosangaisan/fy2016/pdf/02_2.pdf

Ministry of the Environment, Japan (12 April 2013) 'Japan's climate change policies'. *Ministry of the Environment Government of Japan.* Accessed on 5 February, 2016 form www.env.go.jp/en/focus/docs/files/20130412-68.pdf

Ministry of the Environment, Japan (August 2015) 'Heisei 28-nendo Kankyōshō jūtensesaku' [Ministry of Environment's priority policies for fiscal year 2016]. *Ministry of the Environment Government of Japan.* Accessed on 23 February, 2016 from www.env.go.jp/guide/budget/h28/h28juten-1.pdf

Ministry of Health, Labour and Welfare, Japan (2015) *Heisei 27-nenban kōsei rōdō hakusho: jinkō genshō shakai-wo kangaeru* [2015 Edition of Health, Labour and

Welfare white paper: Considering a population decline society]. Accessed on 2 April 2016 from www.mhlw.go.jp/wp/hakusyo/kousei/15/

Ministry of Industry and Information Technology, China (19 May 2015). *Zhongguo zhizao 2025 jiedu zhi yi: Zhongguo zhizao 2025, woguo zhizao qiangguo jianshe de hongwei lantu* [China Manufacturing 2025 Interpretation: China Manufacturing 2025, the grand blueprint for the construction China's manufacturing power construction]. Beijing: Ministry of Industry and Information Technology. Accessed on 3 February 2016 from www.miit.gov.cn/n11293472/n11293832/n11294042/n11481465/16595195.html

Ministry of Internal Affairs and Communications, Japan (20 June 2014) *sumāto japan ICT senryaku* [Smart Japan ICT Strategy]. Accessed on 22 March 2016 from www.soumu.go.jp/menu_news/s-news/02tsushin01_03000264.html

Ministry of Internal Affairs and Communications, Japan (August 2014) *Heisei 27-nendo sōmushō ICT-kankei jūten seisaku* [2015 Ministry of Internal Affairs and Communications major ICT-related policies]. Accessed on 1 April 2016 from www.shinetsu-icc.jp/documents/ictsesaku2015.pdf

Ministry of Internal Affairs and Communications, Japan (September 2014) *robotto kakumei jitsugen kaigi (dai-ikkai)* [Conference on the realization of a robot revolution (Number One)]. Accessed on 23 March 2016 from www.kantei.go.jp/jp/singi/robot/dai1/siryou3–4.pdf

Ministry of Internal Affairs and Communications, Japan (August 2015) *Heisei 28-nendo sōmu-shō shokan yosan gaisan yōkyū no gaiyō* [Budget under the jursidiction of the Ministry of Internal Affairs and Communications for fiscal year 2016: Overview of budgetary approporiations request]. Accessed on 2 February 2016 from www.soumu.go.jp/main_content/000374581.pdf

Ministry of Land, Infrastructure, Transport, and Tourism, Japan (2014) 'Dōro-hō-tō no ichibu o kaisei suru hōritsu' no gaiyō ni tsuite' [On the overview of the 'Act for the Partial Revision of the Road Law']. *Highway Industry Development Organization*, Dōro gyōsei seminā, 2014.08: 1–4. Accessed on 7 March 201 6 from www.hido.or.jp/14gyousei_backnumber/2014data/1408/1408douro-hou_kaisei.pdf

Ministry of Land, Infrastructure, Transport, and Tourism, Japan (n.d.) *The Telework Guidebook kigyō-no tame-no terewāku dōnyū/unyō gaidobukku'-no gaiyō* [Summary of 'The Telework Guidebook–Guidebook of introduction and operation of telework for businesses']. Accessed on 14 May 2016 from www.mlit.go.jp/crd/daisei/telework/guidebook/guidebook_gaiyou.html

Ministry of Land, Infrastructure, Transport, and Tourism, Japan (31 December 2015a) *Sumātointāchenji kaitsū kasho: ichiranpyō* [Smart interchange opening points: A list]. Accessed on 6 March 2016 from www.mlit.go.jp/road/sisaku/smart_ic/donyu.pdf

Ministry of Land, Infrastructure, Transport, and Tourism, Japan (31 December 2015b) *Sumātointāchenji jigyō-chū kasho* [Locations of smart interchanges in business]. Accessed on 6 March 2016 from www.mlit.go.jp/road/sisaku/smart_ic/jigyo.pdf

Mizuho (2016) 'Mizuho sangyō chōsa – Tokushū: Sekai no chōryū to Nihon sangyō no shōrai-zō – gurōbaru shakai no paradaimushifuto to Nihon no shinro' [Mizuho industry research – Special Issue: world trends and future vision of Japanese industry – Global society paradigm shift and Japan's direction]. OneMizuho, *Mizuho* 54(1). Accessed on 21 March 2016 from www.mizuhobank.co.jp/corporate/bizinfo/industry/sangyou/m1054.html

Mizutani, Fumitoshi, and Shuji Uranishi (2006) 'Privatization of the Japan Highway Public Corporation: Policy assessment'. *Paper for the 46th Congress for the European Regional Science Association.* Accessed on 7 March 2016 form www-sre.wu-wien. ac.at/ersa/ersaconfs/ersa06/papers/226.pdf

Nagasaka, Toshinari (2013) 'The Great East Japan Earthquake and issues of risk governance and risk communication in complex and multiple LPHC type of disasters – and report from the society for risk analysis-Japan,' in Saburo Ikeda and Yasunobu Maeda (eds) *Emerging issues learned from the 3.11 disaster as multiple events of earthquake, tsunami and Fukushima nuclear accident.* The Committee of the Great East Japan Disaster, Society for Risk Analysis, SRA Japan, pp. 7–10. Accessed on 10 January 2016 from www.sra-japan.jp/cms/uploads/311Booklet.pdf

Nagashima, Miuri, Glenn D. Hook, and Piers R. Williamson (2015) *kakusan-suru risuku-no seijisei: soto-naru zashi; uchi-naru zashi* [*The Politics of the Spread of Risk: Internal and External Observations*]. Nara: Kizasu shobō.

National Diet (n.d.) 'Kokkai kaigi-roku kensaku shisutemu' [National Diet proceedings search engine]. *National Diet Library.* Accessed on 20 March 2016 from http://kokkai.ndl.go.jp/

National Energy Administration of China National Energy Administration of China (27 July 2015) 'Guojia nengyuan ju fabu zhineng dianneng biao xilie biaozhun' [National Energy Administration issues smart meter standard series]. *Zhongguo zhengfu wang.* Accessed on 13 May 2016 from www.gov.cn/xinwen/2015–07/27/content_2903117.htm

Nicholls, Paul, Tom Goerke, and Andrew Rohl (30 July 2015) 'Cisco internet of everything innovation centre'. *Curtin University.* Accessed on 6 February 2016 from http://research.curtin.edu.au/about/institutes-centres/cisco-internet-of-everything-innovation-centre

Nicholls, Sean (8 March 2016) 'Households face buying $600 smart meter to avoid electricity "bill shock"'. *The Sydney Morning Herald.* Accessed on 9 March 2016 from www.smh.com.au/nsw/households-face-buying-600-smart-meter-to-avoid-electricity-bill-shock-20160308-gndf37.html

Nihon Iryō Kenkyū Kaihatsu Kikō (15 September 2015) 'Shukuhaku-gata shin hoken shidō (sumāto raifu sutei) jigyō-no gaiyō-to shinchoku jōkyō-ni tsuite' [Accommodation-type health guidance (smart life stay): A business overview and state of progress]. *Ministry of Health, Labour and Welfare.* Accessed on 17 March 2016 from www.mhlw.go.jp/file/06-Seisakujouhou-10900000-Kenkoukyoku/20150915siryou1.pdf

Nikkei Shimbun (3 April 2015) 'sumātokā-toha' [What's a smart car?]. *Nikkei Inc.*

Nikkei Shimbun (17 June 2015) 'Hassōdenbunri-o kettei, kaiseiden jihō-ga seiritsu sannyū kyōsō unagasu' [Amended Electricity Business Act established, with decision to separate power generation and transmission – encourages participation and competition]. *Nikkei Inc.* Accessed on 11 March 2016 from www.nikkei.com/article/DGXLNSE2INK01_X10C15A6000000/

Nikkei Shimbun (17 July 2015) 'wāku sutairu kaikaku-ha inobēshon-wo umidasu-no ka?' [Will reform to work style produce innovation?]. *Nikkei Inc.* Accessed on 30 April 2016 from www.nikkeibp.co.jp/atcl/column/15/hc2015/071700023/?rt=nocnt

Nippon Denshin Denwa (2005) 'nettowākurobotto-ni tsuite oshiete kudasai' [Please teach me about network robots]. *NTT jigutsu jānaru* 17(5): 68–69. Accessed on 23 March 2016 from www.ntt.co.jp/journal/0505/

Nishio, Katsuhito (2012) 'Wākuraifubaransu kōjō-to jigyō keizoku keikaku-wo ryōritsusaseru terewāku-no yūkōsei-no kenshō' [Verification of efficacy of telework which balances a business continuity plan with work-life-balance improvement]. University of Hyogo, *Shō dai bijinesurebyū* 2(2): 113–124. Accessed on 1 April 2016 from www.u-hyogo.ac.jp/mba/SBR/2–2.html

Nozawa, Tetsuo (21 April 2011) 'Sofutobanku-no son-shachō, shizenenerugīzaidan-o setsuritsu-e' [President of Softbank, Son, to establish the Japan Renewable Energy Foundation]. *Nikkei Keizai Shimbun*, Nikkei Inc. Accessed on 2 February 2016 from www.nikkei.com/article/DGXNASFK2100O_R20C11A4000000/

Nuclear Energy Institute (2015) *Small Reactor Designs*. Washington, DC: Nuclear Energy Institute. Accessed on 23 January 2016 from www.nei. org/Issues-Policy/New-Nuclear-Energy-Facilities/Small-Reactor-Designs

Oi, Takashi (10 September 2012) 'Privatization of Japan's expressway company and its financing scheme'. Presentation given at the *IBTTA 80th Annual Meeting and Exhibition, International Bridge, Tunnel and Turnpike Association*. Accessed on 5 March 2016 from http://ibtta.org/sites/default/files/Oi_Takashi.pdf

Okiden (30 April 2014) 'sumātomētānado-no seibi keikaku-ni tsuite' [On the plan to development smart meters etc.]. *The Okinawa Electric Power Company, Incorporated*. Accessed on 10 March 2016 from www.okiden.co.jp/shared/pdf/whats_new/2014/140430_02.pdf

Owen, Gill and Judith War (2006) 'Smart meters: Commercial policy and regulator drivers'. *Sustainability First*, London. Accessed on 9 March 2016 from www. ofgem.gov.uk/ofgem-publications/42078/13383-sustainabilityfirstresponse.pdf

Phuangpornpitak, Napaporn and S. Tia (2013) 'Opportunities and challenges of integrating renewable energy in smart grid system'. *Energy Procedia* 34: 282–290.

Pope, Chris G. and Rob Fahey (in progress) 'The transmutation of self-responsibility in the Japanese news media: From buzzword to social norm'.

PR Newswire (19 March 2015) 'Australia – smart cities – people, transport, cars, buildings: Big data and open governments will spur developments in smart cities'. *PR Newswire*. Accessed on 6 February 2016 from www.prnewswire.com/news-releases/ australia – smart-cities – people-transport-cars-buildings-300053487.html

Rignot, Eric, Jeremie Mouginot, Mathieu Morlighem, Helene Seroussi, and Bernd Scheuchl (2014) 'Widespread, rapid grounding line retreat of Pine Island, Thwaites, Smith, and Kohler glaciers, West Antarctica, from 1992 to 2011'. *Geophysical Research Letters* 41: 3502–3509. doi:10.1002/2014GL060140

Rikuden (n.d.) 'sumātomētā dōnyū-ni tsuite' [On smart meters]. *Hokuriku Electric Power Company*. Accessed on 10 March 2016 from www.rikuden.co.jp/info/ smartmeter.html

Rist, Gilbert (2010) 'Development as a buzzword', in Andrea Cornwall and Deborah Eade (eds) *Deconstructing Development Discourse: Buzzwords and Fuzzwords*. Warwickshire, UK: Oxfam and Practical Action Publishing, pp. 19–28. Accessed on 25 March 2016 from www.guystanding.com/files/documents/Deconstructing-development-buzzwords.pdf

Rogai, Sergio (2006) 'ENEL's metering system and telegestore project'. *NARUC Conference*, Washington, 19 February 2006.

Ryūkyū Shimpō (18 January 2016) '< Shasetsu > denryoku zenmen jiyū-ka riyōsha hon'i-no ryōkin taikei-ni' [<Editorial> Full liberalization of electric power – to a user-oriented fee system]. *The Ryukyu Shimpo*. Accessed on 11 March 2016 from http://ryukyushimpo.jp/editorial/entry-205819.html

Ryūkyū Shinpō (29 January 2016) 'ryūkyūshinpō okiden-no ryōkin taikei, 4 tsuki ikō-mo iji kouri jiyū-ka' [Okiden's price system to be maintained after April and retail liberalization]. Accessed on 11 March 2016 from http://ryukyushimpo.jp/news/entry-212499.html

Saitō, Toshikazu (24 February 2014) 'Sumāto raifu purojekuto suishin iinkai' [Smart Life Project Committee], runesansu kaichō-no burogu [Blog of the President of Renaissance Inc.]. *Ameba*, CyberAgent, Inc. Accessed on 1 February 2016 from http://ameblo.jp/re3110/

Sankei News (28 February 2016) '"kōsoku-no anzen-wo mamoru" "dōro kansei sentā" shidō' [Start of the "Road Control Centre" which "protects expressway safety"]. *The Sankei Shimbun & Sankei Digital*. Accessed on 27 April 2016 from www.sankei.com/photo/story/news/160228/sty1602280023-n1.html

Sanseido Web Dictionary (n.d.) 'Sumāto₂ smart' [Smart₂ Smart]. *Sanseido*. Accessed on 11 March 2016 from www.sanseido.net/User/Dic/Index.aspx?TWords=%u30b9%u30de%u30fc%u30c8andst=0andDORDER=andDailyJJ=checkboxandDailyEJ=checkboxandDailyJE=checkbox

Satell, Greg (1 April 2016) 'Why energy storage may be the most important technology in the world right now'. *Forbes*. Accessed on 27 April 2016 from www.forbes.com/sites/gregsatell/2016/04/01/why-energy-storage-may-be-the-most-important-technology-in-the-world-right-now/#32db5d748690

Sato, Akio (2013) 'Teleworking and changing workplaces'. *Japan Labour Review* 10(3): 56–69. Accessed on 2 April 2016 from www.jil.go.jp/english/JLR/backissues/2013.html

Sekizawa, Jun (2013) 'Appropriate risk governance on radionuclide contamination in food in Japan', in Saburo Ikeda and Yasunobu Maeda (eds) *Emerging Issues Learned from the 3.11 Disaster as Multiple Events of Earthquake, Tsunami and Fukushima Nuclear Accident*. The Committee of the Great East Japan Disaster, Society for Risk Analysis, SRA Japan, pp. 31–35. Accessed on 10 January 2016 from www.sra-japan.jp/cms/uploads/311Booklet.pdf

Shahan, Zachary (5 April, 2012) 'Japan's rapid residential solar power growth'. Clean-Technica, *Sustainable Enterprise Media, Inc.* Accessed on 11 March 2016 from http://cleantechnica.com/2012/04/05/japans-rapid-residential-solar-power-growth/

Shaw, Ray (27 February 2016) 'Smart meters, need smart grids, need smart clouds, need smart people'. *iTWire*. Accessed on 9 March 2016 from www.itwire.com/sponsored-announcements/71648-smart-meters-need-smart-grids-need-smart-clouds-need-smart-people.html

Shinohara, Sobee, Yuji Akematsu, Hiroyuki Morikawa, Masatsugu Tsuji (2013) 'Current issues of the Japanese mobile phone market caused by Smartphones'. *Conference paper for the 24th European Regional Conference of the International Telecommunications Society, Florence, Italy, 20–23rd October 2013*. Accessed on 11 March 2016 from www.econstor.eu/bitstream/10419/88528/1/774553014.pdf

Smart City China (20 January, 2016) 'Zhihui chengshi: Jingji zhuanxing kan tixii biange zhi lu' [Smart City: Economic transformation, on the road of systematic change]. *Nanjing Yunchuang Big Data Technology Co. Ltd*. Accessed on 1 February, 2016 from www.smartcitychina.cn/ZhengCeJuJiao/2016–01/6411.html

Smart Life Japan (27 May 2015) 'Sumātou raifu osusume book' [Book recommending a smart life]. *Association for Electric Home Appliances*. Accessed on 5 April 2016 from http://smart-life-japan.jp/pc/about/file/recommend_book.pdf

Smart Life Japan (31 March 2016) 'Soshiki gaiyō' [Organization overview]. *Smartlifejapan Promotion Forum*. Accessed on 4 April 2016 from http://smart-life-japan.jp/company/

Smart Life Project Committee and Ministry of Health, Labour and Welfare, Japan (n.d.) *3-ttsu no akushon + 1* [3 actions + 1]. Accessed on 1 February, 2016 from http://smartlife.go.jp/about/3action/

Smart Life Project Committee and Ministry of Health, Labour and Welfare, Japan (11 December, 2014) *Activities of the Smart Life Project Committee: The Smart Life Project Committee*. Accessed on 1 February, 2016 from http://smartlife.go.jp/about/honbu/2014/12/11

Smart Platinum Society Promotion Council (July 2014) 'sumāto purachina shakai suishin kaigi hōkokusho' [Smart Platinum Society Promotion Council Report]. *Ministry of Internal Affairs and Communications*. Accessed on 29 March 2016 from www.soumu.go.jp/main_content/000303235.pdf

Spegele, Brian (30 March 2016) 'China's state grid envisions global wind-and-sun power network'. *The Wall Street Journal*. Accessed on 31 March 2016 from www.wsj.com/articles/chinas-state-grid-envisions-global-wind-and-sun-power-network-1459348941

State Grid Corporation of China (30 March 2016) '2016 Global Energy Interconnection Conference opens in Beijing'. *PR Newswire*. Accessed on 31 March 2016 from www.prnewswire.com/news-releases/2016-global-energy-interconnection-conference-opens-in-beijing-300243221.html

Strang, David and John W. Meyer (1993) 'Institutional conditions for diffusion'. *Theory and Society* 22: 487–511. Accessed on 31 March 2016 from http://people.soc.cornell.edu/strang/articles/Institutional%20Conditions%20for%20Diffusion.pdf

Strbac, Goran, Chin Kim Gan, Marko Aunedi, Vladimir Stanojevic, Predrag Djapic, Jackravut Dejvises, Pierluigi Mancarella, Adam Hawkes, Danny Pudjianto, Scott Le Vine, John Polak, Dave Openshaw, Steven Burns, Phil West, Dave Brogden, Alan Creighton, Alan Ciaxton (2010) 'Benefits of advanced smart metering for demand response based control of distribution networks'. *Summary Report*, Version 2.0, Imperial College London/Energy Networks Association. Accessed on 9 March 2016 on www.energynetworks.org/assets/files/electricity/futures/smart_meters/Smart_Metering_Benerfits_Summary_ENASEDGImperial_100409.pdf

Takeda, Hiroko (2005) *The Political Economy of Reproduction in Japan: Between Nation-State and Everyday Life*. Oxon, UK: RouteldgeCurzon.

Technische Universität Wien (TUWIEN) (2015a) 'europeansmartcities 4.0 (2015): Home'. *Technische Universität Wien*. Accessed on 7 February 2016 from www.smart-cities.eu/index.php?cid=-1andver=4

Technische Universität Wien (TUWIEN) (2015b) 'europeansmartcities 4.0 (2015): City-Profile: Sheffield (UK)'. *Technische Universität Wien*. Accessed on 7 February 2016 from www.smart-cities.eu/index.php?cid=6andver=4andcity=212

Tochigi Smart Driver (2013) 'tochigi sumāto doraibā-ni tsuite' [About Tochigi Smart Driver]. *Tochigi Smart Driver Project Committees*. Accessed on 27 April 2016 from http://tochigi-sd.jp/info/

Tohoku Electric Power Corporation Inc. (7 January 2015) 'sumātomētā-no secchi kaishi-ni tsuite' [On the installation of smart meters]. *Tohoku-Electric Power Co., Inc.* Accessed on 10 March 2016 from www.tohoku-epco.co.jp/information/1188769_821.html

Tokyo Electric Power Company Inc (2016) 'sumātomētā-ni tsuite' [On smart meters]. *TEPCO Energy Partner Inc.* Accessed on 10 March 2016 from www.tepco.co.jp/smartmeter/index-j.html

Tokyo Electric Power Company Inc. (n.d.) 'sumāto raifu puran' [Smart life plan]. *TEPCO Energy Partner Inc.* Accessed on 4 April 2016 from www.tepco.co.jp/jiyuuka/service/plan/smartlife/index-j.html

Tokyo Smart Driver Project Committees (2011a) *tōkyō sumāto doraibā-ni tsuite* [About Tokyo Smart Driver]. Accessed on 27 April 2016 from www.smartdriver.jp/about/

Tokyo Smart Driver Project Committees (2011b) *sumāto doraibu akushon* [Smart Drive actions]. Accessed on 27 April 2016 from www.smartdriver.jp/action/

Tokyo Smart Driver Project Committees (2011c) *sanka-no shikata* [Ways to join in]. Accessed on 27 April 2016 from www.smartdriver.jp/entry/

Tokyo Smart Driver Project Committees (2011d) *samazama-na katsudō* [Various activities]. Accessed on 27 April 2016 from www.smartdriver.jp/various/

Tokyo Smart Rider (2012) *tōkyō sumāto raidā-ni tsuite* [About Tokyo Smart Rider]. Accessed on 27 April 2016 from http://smartrider.jp/about/

Transmission and Distribution World (9 July 2015) 'Smart electricity meters to total 780 million in 2020 driven by China's roll-out'. *Penton.* Accessed on 13 May 2016 from http://tdworld.com/ami/smart-electricity-meters-total-780-million-2020-driven-china-s-roll-out

Vanolo, Alberto (2014) 'Smartmentality: The smart city as disciplinary strategy'. *Urban Studies* 51(5): 883–898. doi:10.1177/0042098013494427

Wakuta, Yukihiro (2015) 'Ajenda settingu-ni okeru imi nettowāku-to furēmingu: 'Chisanchishō'-o jirei-to shite' [The meaning network and faming in agenda setting: A case study on "Local production for local consumption"]. *Nihon jōhō keiei gakkaishi* 35(3): 4–7. Accessed on 13 February 2016 from http://ci.nii.ac.jp/naid/110009976294

Weblio (n.d.) 'Sumāto' [Smart]. *Weblio.* Accessed on 11 March 2016 from http://thesaurus.weblio.jp/content/%E3%82%B9%E3%83%9E%E3%83%BC%E3%83%88

Williamson, Piers (2014) 'Demystifying the official discourse on childhood thyroid cancer in Fukushima'. *The Asia-Pacific Journal* 12(49): Number 2. Accessed on 10 January, 2016 from http://apjjf.org/2014/12/49/Piers-Williamson/4232.html

Winner, Langdon (1980) 'Do artifacts have politics?'. *Daedalus* 109(1): 121–136.

World Bank (2016) *World Development Report 2016: Digital Dividends. World Development Report.* Washington, DC: World Bank Group. Accessed on 9 February 2016 from www.worldbank.org/en/publication/wdr2016

Yamamoto, Hiromi (2003) 'New public management: Japan's practice'. *Institute for International Policy Studies*, Policy Paper 293E January 2003.

Yamanashi Smart Driver (2016) *yamanashi sumāto doraibā-toha* [What is Yamanashi Smart Driver?]. Accessed on 27 April 2016 from http://sdyamanashi.jpn.org/what/

Yonden (25 November, 2014) 'sumātomētā-no secchi kaishi-ni tsuite' [On the start of smart meter installations]. Press Release, *Yonden Shikoku Electric power Co., Inc.* Accessed on 10 March 2016 from www.yonden.co.jp/press/re1411/1186983_2063.html

Zhang, Xue, Wei Pei, Wei Deng, Yan Du, Zhiping Qi, and Zuomin Dong (2015) 'Emerging smart grid technology for mitigating global warming'. *International Journal of Energy Research* 39: 1742–1758. doi:10.1002/er.3296

7 Conclusion

The empirical chapters of this book have sought to answer the question 'whose pollution?' by examining how responsibility for taking action in response to pollution is reported in the Asia-Pacific media, with a specific focus on the sources of the risks and harms posed by atmospheric and other forms of pollution in Australia, China and Japan. Our approach has been informed by different disciplines, theories, methodologies and ways of answering this question – as outlined in Chapter 1 – but the book maintains the common aim of illuminating the sources, risks and responsibilities associated with pollution in these three countries. This final chapter brings together a number of our key findings; highlights the similarities and differences in how responsibility is manifest, given the sources of pollution in the Asia-Pacific; and draws attention to how the responses to specific cases of pollution are instrumentalised in Australia, China and Japan. The aim is to provide a concise summary of our work by addressing four questions: first, what are the sources of the risks, potential harms and harms posed by pollution? Second, how are these reported in the media of the three countries? Third, who is viewed as being responsible for responding to the risks and harms? Finally, how is the response instrumentalised effective?

Sources of risks and harms

The sources of the risks and harms resulting from pollution differ from country to country, but a common thread running through the empirical chapters is how the risks and harms posed to both the environment and the health of citizens are integral to a modern 'risk society'. For modernisation, especially as manifest in the patterns of energy consumption, has produced risks, potential harms and harms to the environment, as well as to the citizens, especially in terms of their everyday life and health.

In the case of Australia, the world's continuing reliance on fossil fuels means that the mining of coal and the transportation of this 'black gold' across the seas, to be burned for heating industrial furnaces and to generate electricity, has polluted the waters around the Great Barrier Reef, a natural resource for

Australia and the globe. While the erosion of the life of the reef is a concrete example of how the risks of pollution from the mining of coal are causing harm to the reef's coral, fish and other life forms, US President Barack Obama's 2014 speech at the University of Queensland highlights the importance of the reef for the present and for the future of the world, a global future where he hopes his daughters and their children are still able to visit the reef '50 years from now' (Obama 2014).

The most important export destination for Australian coal is China. Australia plays a vital role in filling the country's vast energy needs and remains the key source of overseas thermal coal to generate electrical energy as well as to produce steel and cement, despite a recent decline in coal imports. The pollution caused by the burning of coal is the source of risks and harms to the health of the population, as the case study of China clearly demonstrates. Although not the only source of atmospheric pollution in China, as oil as well as dust whipped up by wind and storm add to the amount of pollutants entering the atmosphere, the burning of coal along with vehicle exhaust fumes contribute the major sources of PM2.5 to the atmosphere, a particulate the evidence shows causes harm to the health of Chinese citizens.

But the PM2.5 released into the atmosphere is not simply a source of risks and harms to the health of the citizens of China, as wind carries the particulate over the sea to Japan as well. In the same way as ships transport raw coal across the sea from Australia to China in a process of transferring one of the sources of atmospheric pollution across the borders between these two states, the Upper Westerlies similarly transport PM2.5 over national borders, this time between China and Japan. The transborder flow of PM2.5 heightens the risk of damage to the health of the citizens of Japan, not just China, including cancer, respiratory disease and other illnesses. Due to its geographic location, transborder atmospheric pollution is especially harmful to the residents of the southern island of Kyushu, although the direction and force of the wind means that populations in other parts of the Japanese archipelago can suffer the harm of PM2.5 as well. In this way, the risks and harms posed by pollution reach across state borders, affecting not only the Great Barrier Reef in Australia, but the health of Chinese and Japanese citizens as well. We see here how the three countries are linked together in a transnational chain of risks and harms. In short, the transborder risk of PM2.5 is testimony to how the sovereign territorial boundaries of the state offer scant protection against environmental risks and harms to the population's health.

Atmospheric pollution is one of the sources of harm to which 'smart cities' and 'smart societies' are meant to adapt. As our empirical chapter on 'smart cities' demonstrates, smart initiatives are being introduced in Australia and China as well as in Japan, the main focus of this chapter. The goal of the smart society is to employ technological innovation in a way to maintain economic growth at the same time as the sources of risks and harms produced by pollution are ameliorated.

Media reporting

The media plays a pivotal role in reporting, communicating and embedding the discourse of risk posed by climate change and the variously attributed responsibility at the international, state, market and societal levels. The qualitative and quantitative methods used to investigate the nature of the media reporting on pollution and the environment in the empirical chapters enabled us to illuminate the different ways pollution is addressed. The focus on Australia demonstrates how the media have taken up the impact of coal mining on the health of the Great Barrier Reef, along with deforestation and wider issues of climate change. What is striking about the Australian case is how environmental risks and harms are contested, with environmentalists and sceptics challenging each other's portrayal of the risks and harms posed by climate change. This means a range of voices are present in the media, illustrating how journalists in Australia have at least some leeway in how they report on environmental issues related to pollution and climate change, despite the high concentration of ownership in the media market and the existence of a degree of self-censorship in the newsroom.

This is in sharp contrast to China. It is precisely due to the high degree of control of journalistic reporting on pollution and climate change more generally that our empirical chapter examined the role of documentary film in communicating the risks and harms posed by climate change and the responsibility to respond, rather than focus on an investigation of the print media. The documentary *Under the Dome* has played a critical role in highlighting the risk of and responsibility for atmospheric pollution in China, demonstrating the particular strengths of this medium for shaping attitudes and promoting action. The images seen in the documentary paint a vivid picture of the role of environmental activism in reporting on air pollution and raising awareness among the citizens of how PM2.5 in smog affects health. It undermines the complacent view of atmospheric pollution as nothing more than fog by offering interviews and concrete, everyday examples of how PM2.5 impacts on the nation's health, as well as the way air pollution can put an end to some of the joys of enjoying the natural environment.

The examination of how PM2.5 is reported in Japan demonstrates the role of other news events in shaping the media's coverage of this atmospheric pollutant. As detailed in our chapter on Japan's response to cross-border atmospheric pollution, a range of other news, representing social, political or environmental events over the period from 1998 to 2016, changes the way PM2.5 is addressed in the media. The temporal dimension reveals how reporting on this particulate increases in salience due to different news events, as illustrated in the period up until January 2013 when the media reported on how emissions from factories and automobiles lead to an increase in atmospheric pollution, with attention paid to the situation in China and Japan's own history of pollution. At other times, the pollution in China is at the heart of media reporting, showing how a strong interest in the source of cross-border atmospheric pollution is helping to shape the way the Japanese media report on PM2.5. The coverage of the

Chinese government's response to atmospheric pollution shows the media playing a crucial role in shaping environmental discourse, as the reported lack of commitment to reducing emissions on the part of the government is viewed as evidence of the continued flow of cross-border pollutants to Japan in the future. Sometimes the reporting is prompted by domestic events, as illustrated by the Ministry of the Environment's decision later in 2013 to move forward with a policy based on adaptation to the risks and harms of climate change, prompted by the Fifth Report of the Intergovernmental Panel on Climate Change. This marked the government's greater prioritisation of adaptation to climate change rather than mitigation of pollution.

At the same time, our study of reporting in the Japanese media and Diet debates analysed how 'smart cities' and 'smart societies' are part of the way the government is hoping to move the country away from the risks and harms of pollutants such as PM2.5 to a new, low-carbon economy with reduced emissions. The chapter's detailed tracing of the use of 'smart' indicates how reporting has made the term a buzzword which informs a range of approaches to climate change, including the goal of the Japanese government to make Japan a 'super smart society' before the Tokyo Olympics in 2020. Reporting and Diet debates on 'smart' issues do not directly focus on the risks and harms posed by pollution and climate change more broadly, but rather on how technology can be used to enable society to adapt to climate change and mitigate pollution. 'Smart societies' are able to use a range of smart technologies, such as 'smart meters', as a way to try to reduce energy consumption and smart interchanges on Japanese toll roads as a means to not only cut costs, but to smooth movement along the highway and thereby reduce fuel consumption and emissions.

Responsibility

The question of who takes responsibility in responding to pollution is at the heart of our enquiry. The empirical chapters reveal how the scope and attribution of responsibility for atmospheric and other forms of pollution share similarities and differences across three countries. All three countries respond to climate change and address the risks and harms posed by pollution in the wider global context, not simply the national or local context. This can be seen in the case of Australia, for instance, where the empirical chapter offers insights into how an international organisation helped to change the dynamics of responsibility domestically. The role of UNESCO's World Heritage Committee in bringing the question of responsibility within Australia to international attention is illustrative. In competition with environmental groups seeking to list the Great Barrier Reef as 'in danger,' both the national government and the subnational government of Queensland fought a successful battle to prevent this from occurring. What this suggests is that the reputational risk and potential economic harm to Australia of the reef being branded as 'in danger' by an international organisation lead these two levels of government to take responsibility for action to help to mitigate harm to the reef, as evidenced by the commitment to cut the run-off of pollution. This demonstrates how the question of responsibility can spill over the boundaries of the state, with international organisations contributing to the shape of the domestic response.

In the case of China, the documentary *Under the Dome* shows how differences in the regulatory environment in a country can make the mitigation of environmental risks and harms a major challenge. The point is not so much the lack of a legal framework to deal with issues like the emission of pollutants by factories and vehicles in different parts of the country. State and substate actors have put in place a host of regulations, which set limits to the amount of permissible emissions by both commercial and domestic users. It is rather that, despite the regulatory frameworks, the regulations are not being enforced through the strict application of the law when infringements do occur. Even though the officials working for the Ministry of Environmental Protection operate in a regulatory framework calling for the application of the law against offenders, they often lack the power to bring to justice factories that flaunt environmental regulations. Or, again, a lack of clarity in who takes responsibility for dealing with vehicles producing emissions beyond the legal limit means their production and use continue. In other words, the question of responsibility is embedded in societal and legal structures as well as norms, which differ among Australia, China and Japan.

The Japanese case study of PM2.5 offers empirical evidence to demonstrate how the loci of responsibility to respond to atmospheric pollutants change over time, illuminating the way responsibility includes a range of stakeholders, including the individual. We see how the communication of the risks and potential harms from pollutants and the responses to them involve state, subnational, market and societal actors in a range of activities. The process is one of the state off-loading the responsibility to respond to different actors, starting with the state building a sense of the risk posed by climate change at the subnational, market and societal levels. The discourse of the Japanese state itself may be in response to international actors, such as the reports issued by the Intergovernmental Panel on Climate Change. In the example of the transborder atmospheric flow of PM2.5 from China, subnational political authorities play a pivotal role in monitoring the atmosphere and issuing warnings to the local inhabitants to take precautionary measures in order to mitigate the potential harm when breathing air is harmful to health. Thus, schoolchildren refrain from playing outdoors when the risk posed by PM2.5 is high, shifting the locus of responsibility from the action of the local government in measuring pollutants to the schools and parents to warn children of the risk and potential harm. At the same time, increased risk provides an opportunity for market actors to respond to societal needs by producing face masks for individual use in filtering out harmful pollutants when breathing. In this way, responsibility cascades down from the state to the citizen, who comes to exercise 'self-responsibility' by taking precautionary measures against the risks and harms posed by PM2.5 and other pollutants.

Similarly, as far as 'smart cities' are concerned, the Japanese state has sought to strengthen the norm of the autonomous individual and promote the entrepreneurial-self in a way to inculcate 'self-responsibility' in the citizenry of the 'smart society'. The smart citizen is adept at monitoring the energy bills produced by smart meters and uses gas and electricity more efficiently, that is, in an environmentally friendly way. Other smart citizens are involved in the production of energy, as residential producers of solar energy deploy their own energy at home and sell on any surplus to larger operators through feed-in tariffs. In order to promote safe as well as smart driving, moreover, drivers in Tokyo can access market actors who provide training to become a 'smart driver', that is, one who not only takes care not to cause an accident, but one who is also able to reduce fuel consumption when driving and thereby act in an environmentally responsible way. But in a smart society not all workers travel: telework serves to cut down on the number of workers needing to commute, as the 'smart worker' can carry out employment duties from home, thereby reducing the risks to the environment from atmospheric and other pollutants. In this way, the smart worker is a self-responsible individual who considers how to reduce energy consumption and thereby contribute to the reduction in pollution, with the norm of being self-responsible for the environment inculcated as part of becoming 'smart'.

Instrumentalisation

We have seen above how a range of actors take responsibility in responding to pollution in Australia, China and Japan, with international organisations, state, substate, market and societal actors involved in a range of activities related to mitigating the harm of pollution, as well as more broadly adapting to the risks of climate change. The empirical case studies demonstrate how these different actors instrumentalise their responsibility, as seen in their tendency to adopt a 'problem-solving' approach to practice. That is, those with attributed responsibility for addressing pollution instrumentalise their responsibility, accepting 'the prevailing social and power relationships and the institutions into which they are organised, as the given framework for action. The general aim of problem-solving is to make these relationships and institutions work smoothly by dealing effectively with particular sources of trouble' (Cox 1981: 128–129). The particular source of trouble these actors face is how to combat pollution and other forms of climate change, but these attempts are mainly carried out within the current framework for action, using a combination of mitigation and adaptation to address the issues faced.

But such a mode of addressing these issues may need to be complemented by another approach, that is, a 'critical-theory' approach to pollution and climate change more broadly. In contrast to the 'problem-solving' approach, critical theory 'does not take institutions and social and power relations for granted but calls them into question by concerning itself with their origins and how and whether they might be in the process of changing' (Cox 1981: 129). As we have seen, the responsibility for addressing pollution has been instrumentalised through a problem-solving approach based on risk management rather than a fundamental challenge calling for the rethinking of contemporary institutions and social/power relations. This study of 'whose pollution?' suggests that it is in combining these two approaches that we may find the most effective way to take on responsibility in response to pollution and climate change in the Asia-Pacific.

References

Cox, Robert W. (1981) 'Social forces, states and world orders: Beyond international relations theory'. *Millennium–Journal of International Studies* 10: 126–155.

Obama, Barack (2014) *Remarks by President Obama at the University of Queensland*. Accessed on 28 September 2016 from www.whitehouse.gov/the-press-office/2014/11/15/remarks-president-obama-university-queensland

Index

Abbott, Tony 22
ABC 23
Abenomics 133
Abe Shinzō 36, 101–2, 113–15, 133, 153
Aboriginal communities 21, 52
activism *see* environmental activism
activist journalism 23–4, 75–6, 82, 89–93
Adani Group 50, 55–6, 59
adaptation policy 12, 101–5, 113, 132–3
Age (Melbourne) 20–1
air pollution in China: action plan for 113; air filter masks and 78; economic growth of China and 76–8; environmental activism and 80–2; environmental movement against 80–2; environmental responsibility attribution to 83–7; environmental risk and 83–7; media and 80–2; as most visible pollution 77; overview 9, 74–6; PM2.5 pollutant levels and 27, 77, 84, 128; public contestation over 78–80; *see also Under the Dome*
APEC summit (2014) 80, 84, 90, 115
Arase Dam removal 38
Arup 149
Asahi Shimbun case 34
ASEAN 50
Ashio copper mine 18, 31–2
Asian Super Grid 150–1
Asia-Pacific Economic Cooperation (APEC) summit (2014) 80, 84, 90, 115
Asia-Pacific region: Australia 18–24; China 24–31; climate change and, constant response to 98; geographical focus of 13; Japan 31–7; media-politics-environment relations in 3; overview 17–18, 37–9
associated word pairs 7, 10

Association of Southeast Asian Nations (ASEAN) 50
atmospheric pollution 189
Australia: Carmichael mine in 50–1, 55; climate change and 22, 38; environmental activism in 20, 22, 56–7; environmental politics of 21–2; federal election of 2013 in 23; history shaping 46; indigenous communities in 21–2; Industrial Revolution and 46; Minerals Council of Australia in 49; natural resources in, access to 22, 48; physical features of 18; public opinion on environmental issues in 24, 38; regional overview 18–24; smart cities in 149–50; smart meter in 161; *see also* Australia's media; coal pollution in Australia; Great Barrier Reef
Australian Broadcasting Corporation (ABC) 23
Australian Survey of Social Attitudes (2013) 24
Australian, The 23, 58
Australian Year Book 47
Australia's media: commercial television and radio 23; environmental activism and 23–4; environmental issues in 22–3; environmental movements in 17, 19–20; environmental risk in 17; Franklin River damming in 17, 20–1; human-induced environmental harm in 19; love of nature in 19; natural disasters in 18–19; news market 23–4; Oxford University's study of 23; protecting landscapes from development in 19–20; *see also* coal pollution in Australia; Great Barrier Reef
automatic data coding and processing techniques 5

BBC 92
Bellamy, David 20
Ben & Jerry's ice cream 56–7
"big four" pollution cases in Japan 18, 32–4
Bishop, Julie 56
British Broadcasting Corporation (BBC) 92
buzzwords 12

CA 106, 108
cadmium poisoning in Toyama prefecture 33
Calhoun, Craig 78
Cao Xianghong 86
carbon capture and storage measure 52
carbon dioxide emissions 27, 51, 53, 114
Carmichael mine 50–1, 55
Carson, Rachel 82
case studies: countries at heart of 37–8; deviant 76; function of 5–6
Castells, Manuel 53
CCTV 80, 82, 93
CDA 12, 110
Centre for Computer Corpus Research on Language of the University of Lancaster (UK) (UCREL) 60
Chai Jing 75, 82–86, 88, 90–3; *see also Under the Dome*
Chen Jining 87
Chen Zhu 83
Chernobyl nuclear accident 35
China: air filter masks in 78; All-China Environment Federation in 88; Australia's coal production and 189; cancer rates in 27; Central Commission for Discipline Inspection in 88–9; China Stone Coal Project in 50–1; climate change and 27, 38; coal production and pollution in 49–51, 76; depoliticized politics in 81–2; "ecological civilization" and 28; economic growth of 17–18, 38–9, 76–8; environmental activism in 9, 26, 29–30, 79–82, 86–7; environmental law in 28, 78, 80–2, 85, 130; environmental movements in 78–82; environmental protection in 26–8; environmental risk in 25–7, 38–9, 75–8; globalization and 27–8, 38–9; "greening of the state" in 78; "green public sphere" in 74, 78; Ministry of Environmental Protection in 85–6; Ministry of Industry and Information Technology in 151; Ministry of Science and Technology in 90; modernisation of 17–18, 25; National Development and Reform Commission 2015 report in 88; National Development and Reform Commission report (2015) in 88; National Energy Administration in 151; natural disasters in, dealing with 25; non-governmental organisations in 79–80, 86–7; physical features of 24–5; PM2.5 pollution in 27, 77, 84, 128; political system and environmental politics in 28–9, 74; propaganda system in 30, 75; public attitude toward environmental issues in 25, 77; Red Alert in 130; regional overview 24–31; renewable energy system in 151; smart cities in 150–2; smart meter in 161; South-North Water Transfer Project in 26; State Council Leading Office for Environmental Protection in 28, 79; State Environmental Protection Administration in 28, 79; State Grid Corporation of China in 151; sustainable development in 26; Three Gorges Dam in 26, 38; water pollution in 27, 77; *see also* air pollution in China; China's media; *Under the Dome*
China Central Television (CCTV) 80, 82, 93
China Mobile 151
China's media: activist journalism in 92; commercial television 80, 82, 93; environmental activism and 80–2; environmental risk in 18, 29–30; Lake Taihu algal bloom in 30; natural disasters in 18; Sichuan Province earthquake in 30; state control of 30–1; technological advancements and 30; "Xiamen PX" protests in 30; *see also* air pollution in China; *Under the Dome*
China Stone Coal Project 50–1
China Telecom 151
China Unicom 151
Chomsky, Noam 23
Cisco Systems 150
clean air measures 52
Climate-ADAPT strategy in European Union 103
climate change: adaptation policy and 102, 113; Asia Pacific region's

response to, constant 98; Australia
and 22, 38; China and 27, 38;
critical-theory approach to 193–4;
Japan and 38, 102, 113; problem-
solving approach to 193; transnational
pollution and 98; United Nations
Framework on 102
Climate Change Adaptation and
Mitigation for Sustainable Risk
Management report 101–2
coal pollution in Australia: coal as
pollutant and 51–2; future coal
production and 49–51, 188–9; global
warming and 51; Great Barrier Reef
and 52–3, 66, 188–9; past coal
production and 46–9; as source of
risks and harms 188–9; transnational
pollution and 45
coding rules 106, 109, 121
Conference of the Parties 15 (COP15) 23
Conference of the Parties 21 (COP21)
102, 115
Conservation Action Trust 56
co-occurrence network analysis in
PM2.5 air pollution in Japan:
coding rules and 106, 125–6, *126*;
description of 108–9; periods of news
articles on, historical 100, 110, *111*,
112–16, *116*, 131; purpose of 108;
shifting semantic contents in media
and 116–21, *117*, *121–2*; temporal
shift in locus of responsibility and
127, 131
co-occurrences of words *see*
co-occurrence network analysis
Cook, James 46
COP15 23
COP21 102, 115
coral mortality *see* Great Barrier Reef
corpus data-based approach to media
data analysis 6–8, 60
corpus data-driven approach to media
data analysis 6–8, 60
correspondence analysis (CA) 106, 108
Courier Mail (Brisbane) 57
critical discourse analysis (CDA) 12, 110
critical-theory approach to climate
change 193–4
cross-sectoral interactions 3, 8–9, 29, 53
customer orientation 159

Daily Mail 55
data mining methods 5–6, 10–11, *10*,
60–1

descriptive statistics 106
deviant case studies 76
discourse analysis *see* PM2.5 air
pollution in Japan; semantic analysis
Doha meetings (2014 and 2015) 57–8
Dunlop, Ian 47–8

earthquakes 18, 30, 34, 36–7, 152
Edmonds, Richard Louis 81
EFA 10
empirical research: environmental
responsibility attribution and
191–3; findings of 13–14, 188–94;
instrumentalisation of responsibility
and 193–4; interdisciplinary
collaboration trend and 13–14; media
reporting and 190–1; multisectoral
interaction models and 8; purpose of
188; sources of risks and harms and
188–9; *see also* Great Barrier Reef;
PM2.5 air pollution in Japan; smart
society in Japan; *Under the Dome*
ENGOs 81
Environment Agency in Japan 33
environmental activism: air pollution in
China and 80–2; Ashio copper mine
and 32; in Australia 20, 22, 56–7;
Australia's media and 23–4; in China
9, 26, 29–30, 79–82, 86–7; China's
media and 80–2; Great Barrier
Reef and 17, 20, 56; in Japan 32;
media 75–6, 80–2, 89–93; by non-
governmental organisations 79–80,
86–7; not-in-my-backyard 79; Three
Gorges Dam and 26; *Under the Dome*
and 75–6, 190
environmental law: in China 28, 78,
80–2, 85, 130; in Japan 33–4,
113–14, 127–8, 159, 161–2
environmental movements: in Australia's
media 17, 19–20; in China 78–82; in
Japan's media 32, 34
environmental NGOs (ENGOs) 81
Environmental Protection Agency
(EPA) of United States 51, 92
environmental responsibility
attribution: to air pollution in China
83–7; constructing 4; empirical
research and 191–3; to Great Barrier
Reef damage 54–66; literature
within media on 4; within media 4;
off-loading in Japan 100; to PM2.5
air pollution in Japan 105–31, 192;
preferred discourse on 99; process of

100–1, 105; qualitative analysis of, to Great Barrier Reef 54–9, 62, 68; semantic analysis of, to Great Barrier Reef 59–66, *61–3*, 68; shared 102; in *Under the Dome* 83–7, 190
environmental risk: of air pollution in China 83 7; in Australia's media 17; in China 25–7, 38–9, 75–8; in China's media 18, 29–30; to Great Barrier Reef 17, 23, 45, 53; interest and action on 2; in Japan 39, 99; in Japan's media 18; of PM2.5 air pollution in Japan 98, 104–31; preferred discourse on 99; as tool of governance 104; transnational flow of political information about 3; in *Under the Dome* 83–7, 190, 192
environmental social innovation and engineering 9, 12
EPA (United States) 51, 92
Ernst & Young Institute 152
EU 102–3
Europe 51, 149
European Smart City Model 149
European Union (EU) 102–3
exploratory factor analysis (EFA) 10

Fairfax Media Limited 23
Federation of Electric Power Companies (FEPC) in Japan 162
"Fight for the Reef" campaign 56
Flannery, Tim 54–5
Foxwell-Norton, Kerrie 57
framing analysis 7
Franklin River damming 17, 20–1
Fukushima Daiichi nuclear plant disaster 34–7, 152

GBRMPA 53
GHG emissions 51, 101–2
globalised research in communication field 1–5, 14
Global Times (China) 88
global warming 51; *see also* climate change
Goenka, Debi 56
GONGOs 29, 38, 79, 88
Gore, Al 82, 89
government-led model 9
government-organised NGOs (GONGOs) 29, 38, 79, 88
graphical analytical techniques 7, 11–12, *11*

Great Barrier Reef: approaches needed in studying 66–7; Australian identity and 52–3; in Australia's media, focus of 68–9; coal pollution in Australia and 52–3, 66, 188–9; conclusions from studying 66–9; coral mortality and 53; environmental activism and 17, 20, 56; environmental debate about, prominence of 21; environmental risk posed to 17, 23, 45, 53; "Fight for the Reef" campaign and 56; importance of 52; keyword searches in studying 62, 68; media reporting conclusions 190; overview of studying 3, 52–3; policymaking and, lack of effective 67; qualitative analysis of environmental responsibility attribution 54–9, 62, 68; quantitative analysis of 68; "Reef Scoop Tour" and 56–7; semantic analysis of responsibility attribution 59–66, *61–3*, 68; as symbol of politics and climate change 22; Yasi cyclone and 53
Great Barrier Reef Marine Park Authority (GBRMPA) 53
Great East Japan earthquake and tsunami and nuclear crisis 18, 34, 36–7, 152
green bans 20
green development 151
greenhouse gas (GHGs) emissions 51, 101–2
Greenpeace 55
"greenspeak" 78
Guardian (Australian edition) 22, 54, 56, 67

heatmap 109, 122–5, *124*
Herman, Edward S. 23
high-efficiency super or ultra-supercritical plants 52
histograms 7, 106
Ho, Peter 81
Hosono Gōshi 36
Huffington Post 87
Hu Jintao 28
Hutchins, Brett 3, 89

ICT 13, 148, 150, 163–4, 166
IEA 52
Inconvenient Truth, An (Gore) 82, 89
India coal production 51

Indonesia 49
information communication technology
(ICT) 13, 148, 150, 163–4, 166
intelligent transport system (ITS) 158
interdisciplinary research 1–5, 9, 14,
99–100, 188
Intergovernmental Panel on Climate
Change (IPCC) 101–3, 113–14,
191–2
International Energy Agency (IEA) 52
IPCC 101–3, 113–14, 191–2
IPCC 38 (2014) 103
ITS 158
ivory ban 28

Jaccard coefficient 109
Janakaraj, Jeyakumar 56
Japan: adaptation policy in 12, 101–5,
113, 132–3; Arase Dam removal
in 38; Australia's natural resources
and, access to 22, 48; "big four"
pollution cases in 18, 32–4; cadmium
poisoning in Toyama prefecture
33; climate change and 38, 102,
113; coal production in Australia
and 48, 50; copper mining in 31;
"development-for-import" policy
in 48; economic revitalization
program in 146; electricity rates
in 168; Environment Agency in
33; environmental activism in 32;
environmental law in 33–4, 113–14,
127–8, 159, 161–2; environmental
protests in 38; environmental risk
in 39, 99; Federation of Electric
Power Companies in 162; Fomenting
Rebellion Act in 32; Fourth
Strategic (Fundamental) Energy
Plan in 114; Fukushima Daiichi
nuclear plant disaster in 34–7, 152;
Headquarters for Countermeasures
for Environmental Pollution in 32–3;
Healthcare Reform Act in 166; health
initiatives in 166–7; history shaping
31; "hydrogen economy" in 153;
industrial development of 31; Japan
Association of Corporate Executives
in 166; Japan Expressway Holding
and Debt Repayment Agency in
159–60; Japan Highway Public
Corporation in 159; Kawamata
incident in 32; Labour Standards
Law in 166; methylmercury

poisoning in Kumamoto and Niigata
prefectures 33–4; Minamata disease
in Kumamoto and Niigata prefectures
33–4; Ministry of Economy, Trade
and Industry in 168; Ministry of
Environment in 102–3, 113–14, 127;
modernisation of 18; multinational
corporations in 39; NEXCO East in
163; nuclear power in 35–6, *35*; off-
loading environmental responsibility
in 100; policy communication process
in 99; Proactive Foreign Policy
Strategy against Global Warming in
102; public opinion on environmental
issues in 35–6, *35*, 38; regional
overview 31–7; smart cities in 152–3;
smart driver campaign in 164; Smart
Life Project Committee in 167; Smart
Life Project in 166–9; smart meter
in 161–2; smart platinum society
in 164–6; Smart Platinum Society
Promotion Council in 164–5; smart
work initiatives in 165–6; sulphur
dioxide pollution in Mie prefecture
34; super smart society goal of 191;
telework initiatives in 165; Yokkaichi
petrochemical complex pollution in
34; *see also* Japan's media; PM2.5 air
pollution in Japan; smart society in
Japan
Japan's media: adaptation policy in 12,
101–5, 113, 132–3; *Asahi Shimbun*
case in 34; Ashio copper mine in 18,
31–2; "big four" pollution cases in
18, 34–7; commercial television and
radio 36; environmental movement
in 32, 34; environmental risk in 18;
Fukushima Daiichi nuclear plant
disaster in 18, 34, 36–7; Great East
Japan earthquake, tsunami and
nuclear crisis of 18, 34, 36–7; as
"information cartels" 37; paraphrases
in, web of 109; policy communication
process in 99; transnational pollution
in 100; trust in 36–7; *see also* PM2.5
air pollution in Japan; smart society
in Japan

Kasperson, Roger E. 103–4
Kawamata incident 32
keyword searches: creating 10; for
Great Barrier Reef damage 62, 68;
for PM2.5 air pollution in Japan 105,

109, 116–22, *116–17, 121–2*; textual analysis and 10
Kyoto protocol 26
Kyushu Electric Power Company 36

Lake Taihu algal bloom 30
Lancaster semantic analysis tool 60
Lane, Marcus 57
Lester, Libby 3, 89
Li Keqiang 87, 130
Liu Zhenya 151
Longman Lexicon of Contemporary English (McArthur) 60

McArthur, Tom 60
Mao Zedong 25
mapping of data points 7
Market Forces financial activists 50
market orientation 162
Markov chain Monte-Carlo (MCMC) 110
Masayoshi Son 152
MCMC 110
MDS 108
media: air pollution in China and 80–2; corpus data-based approach to analysing 6–8, 60; corpus data-driven approach to analysing 6–8, 60; empirical research and reporting of 190–1; environmental activism and 75–6, 80–2, 89–93; environmental reporters and, first 20; environmental responsibility attribution within 4; linguistically orientated approaches to analysing 6, 9; PM2.5 air pollution in Japan topics and, shifting 110, *111*, 112; roles in communication of environmental harms and politics 4–5; social 30, 79; transnational discourses within 45; *see also* Australia's media; China's media; Japan's media
media activism 23–4, 75–6, 82, 89–93; *see also Under the Dome*
media discourse *see* Australia's media; China's media; Japan's media; textual analysis
media-led and popular science-supported model 9
media-politics-environment relations in Asia-Pacific 3
mediated discourse transition 12
methodological innovation: corpus data-based approach 6–8; corpus data-driven approach 6–8; cross-sectoral interactions 3, 8–9; data

mining methods 5–6, 10–11, *10*; framing analysis 7; graphical analytical techniques 7, 11–12, *11*; multisectoral interaction 8–13; practice-focused analysis 9; qualitative analysis of contrastive textual patterns 11–12, *11*; textual analysis 7, 10–12, *10–11*; visualisation 7
methylmercury poisoning in Kumamoto and Niigata prefectures 33–4
Milligan, Spike 20
Minerals Council of Australia 49
Minimata disease 33–4
Minobe Ryōkichi 32
Montreal protocol 26
multidimensional scaling (MDS) 108
multisectoral interactions 8–13
Murdoch, Rupert 23

National Energy Administration (NEA) of China 151
network robots 166
"network society" 53
New Public Management (NPM) 153, 159
News Corp 23
NEXCO East 163
NGOs 29, 76, 78–80, 86–7
non-governmental organisations (NGOs) 29, 76, 78–80, 86–7
not-in-my-backyard (NIMBY) movement 79
NPM 153, 159
nuclear accidents 34–7, 152

Obama, Barack 56, 189
O'Gorman, Dermot 57

Palaszczuk, Annastacia 59
part-of-speech (POS) scheme 110
PCA 10
Pearson's principal component analysis 108
People's Daily newspaper 90, 92
People's Daily website 87–8
Perry, Elizabeth 81–2
Perry, Matthew 31
PFIs 153
PM2.5 air pollution in Japan: action plan for 113–14; adaptation policy and 101–5; analytical methods of studying 106; coded words in studying, English translation of 134–41; coding rules in studying

109, 121; coding text-internal relationships of concepts in studying 121–7, *123–4*; correspondence analysis of 106, 108; critical discourse analysis of 110; data collection in studying 105–6, *107*; descriptive statistics in studying 106; discourse analysis of 12, 99, 106, 110, 126–7; edges plotted in studying 117, *117*, 127; environmental responsibility attribution to 105–31, 192; environmental risk of 98, 104–31; heatmap in studying 109, 122–5, *124*; histogram in studying 106; historical analysis of 112–6, *112*; homogeneity of text in studying 106, *107*, 112, *112*; interdisciplinary analysis of 99–100; keyword searches in studying 105, 109, 116–22, *116–17*, *121–2*; locus of responsibility for, shift in 127–31; multidimensional scaling in studying 108; news sources about, variety of 106, *107*; overview of study 8, 98–101; periods of news articles and, historical 100, 110, *111*, 112–16, *116*, 131; preferred discourse and 99; publications on, number of 12, 110, *111*, 112; qualitative analysis of 106, 110, 132; quantitative analysis of 106, 110, 132; risk discourse and 99; semantic contents in media and, shifting 116–21, *117*, *121–2*; social amplification of risk framework and 103–4; as source of risks and harms 189; suspended particulate matter and 127; temporal shifts in locus of responsibility for 127–31; text analysis tools in studying 110; topics in media, shifting 110, *111*, 112; transnational pollution and 189
POS scheme 110
PPPs 153
practice-focus analysis 9
preferred discourse 99
principal component analysis (PCA) 10
private finance initiatives (PFIs) 153
problem-solving approach to climate change 193
public-private partnerships (PPPs) 153

qualitative analysis: of contrastive textual patterns 11–12, *11*; of Great Barrier Reef environmental responsibility

attribution 54–9, 62, 68; of PM2.5 air pollution in Japan 106, 110, 132; of *Under the Dome* 74–5
quantitative analysis: of Great Barrier Reef 68; of PM2.5 air pollution in Japan 106, 110, 132; of *Under the Dome* 74–5
Queensland Mackay Conservation group court case 55

"Reef Scoop Tour" 56–7
Repnikova, Maria 81
research methodologies *see* methodological innovation
responsibility *see* environmental responsibility attribution
risk *see* environmental risk
risk communication theory 103–5
risk discourse 99
Rudd, Kevin 22

SARF 103–5
SARS 86
self-actualization 13
self-efficiency 164
self-responsibility 12–13, 104, 164
semantic analysis: categories of specific discourse functions and 60, *61*, 62; coding categories 10–11, *10*; component of, key 59; data set for 62–3, *62–3*; defining 59; of Great Barrier Reef environmental responsibility attribution 59–66, *61–3*, 68; high-frequency words in Great Barrier Reef responsibility attribution 63–6; Lancaster semantic analysis tool and 60; methodological considerations for using 66–7; purpose of 59; terminology 59–60
semisupervised data coding and processing techniques 5
severe acute respiratory syndrome (SARS) 86
shared environmental responsibilities 102
Showa Denko factory 33
Sichuan Province earthquake 30
Silent Spring (Carson) 82
Sinopec 86
Sixteenth Tripartite Environment Ministers Meeting (TEMM) (2014) 114
smart cars 158
smart cities: in Australia 149–50; in China 150–2; concept of 148;

defining 152–3; in Europe 149; in Japan 152–3
smart citizens 193
smart initiatives 149, 153
smart innovations 148
smart interchange 156–60
smart life and individual 163–9
smart meter 160–2
smartphone 157–8
smart restructuring 151
smart shrink model 153
smart society in Japan: atmospheric pollution and 189; diffusion of smart initiatives and 147; domains of 148; individuals in, autonomous 193; intelligent transport system and 158; overview 146–9, 169–70; self-efficiency and 164; self-responsibility and 164; *smart* applications in National Diet and 155–8, *155–7*, 163; *smart* as buzzword and 12, 146, 148, 154, 163–70; smart cars and 158; smart categories compiled from National Diet proceedings 171–2; *smart* definition and 154–5; smart driver campaign 164; smart future and, visions of 148, 153–69; smart interchange and 156–60; Smart Life Project and 166–9; Smart Life Project Committee and 167; smart meter and 160–2; Smart Platinum Society Promotion Council and 164–5; smart work initiatives 165–6; textual analysis 163–9
smart work initiatives 165–6
social amplification of risk framework (SARF) 103–5
social media 30, 79
soft journalism 75
Southeast Asia coal production 51
South-North Water Transfer Project 26
SPM 127
sulphur dioxide pollution in Mie prefecture 34
super smart society concept 148; *see also* smart society in Japan
Su Shulin 89
suspended particulate matter (SPM) 127
sustainable development 26
"switching points" 53

Tanaka Shōzō 32
Tasmania 17, 19–21, 46
Technology, Entertainment, Design (TED) talks 75, 82

telework initiatives in Japan 165
TEMM16 (2014) 114
text-internal relationships 100
textual analysis: contrastive patterns, graphical 11–12, *11*; function of 7; semantic coding categories 10–11, *10*; smart society in Japan 163–9; text-internal relationships 100; tools 110; word clustering 7, 11–12, *11*; word occurrence frequencies 10; word pair associations 7, 10
Thiess Peabody Mitsui company 48
three Es (efficiency, effectiveness, economy) 153
Three Gorges Dam 26, 38
Three Mile Island nuclear accident 35
Tōōnippō (Aomori prefectural newspaper) 37
Toshikazu Saitō 167
translation of differences 1–2
transnational pollution: climate change and 98; coal pollution in Australia and 45; by fossil fuels 3–4; in Japan's media 100; *see also* Great Barrier Reef; PM2.5 air pollution in Japan; *Under the Dome*

UCREL Semantic Analysis System (USAS) 60
Under the Dome (Chai Jing): activist journalism and 75–6, 89–93; authorities' reaction to 87–8; beginning of 83; as deviant case study 76; ending of 86–7; environmental activism and 75–6, 190; environmental responsibility attribution in 83–7, 190; environmental risk in 83–7, 190, 192; markets' reaction to 87; media activism and 75–6, 89–93; media reporting conclusions and 190; overview of study 82–3; PM2.5 description in 83; political constraints on media activism and 89–92; purpose of studying 93; qualitative analysis of 74–5; quantitative analysis of 74–5; release of 75, 82; response to 87–9; severe acute respiratory syndrome and 86; soft journalism and 75; TED talk-presentation style of 75, 82
UNESCO 20–1, 53, 57–8, 67, 191
United Nations (UN) Framework on climate change 102

United Nations Educational,
  Scientific and Cultural Organization
  (UNESCO) 20–1, 53, 57–8, 67, 191
United States 28, 35, 51, 92
United Tasmania Group 19
USAS 60

visualisation 7

Waisbord, Silvio 1–2
Wang Tianpu 89
water pollution in China 27, 77
Weblio website 109
WHO 27, 113
World Bank 26–7
World Health Organization (WHO)
  27, 113

World Heritage area 20–1
World Heritage Committee 52, 57–8,
  67, 191

"Xiamen PX" protests 30
Xie Zhenhua 90
Xi Jinping 88, 90

Yang, Guobin 78
Yao Ming 27
Yasi cyclone 53
Yokkaichi petrochemical complex
  pollution 34
Youku website 82

Zhou Shengxian 29, 90
Zhou Yongkang 89